Flight Training Manual, 4th edition

Transport Canada
Aviation

Flight Training Manual
4th edition

gage EDUCATIONAL PUBLISHING COMPANY
A DIVISION OF CANADA PUBLISHING CORPORATION
Vancouver · Calgary · Toronto · London · Halifax

Published by Gage Educational Publishing Company, a division of
Canada Publishing Corporation, and Transport Canada in co-operation
with the department of Public Works and Government Services.

Canadian Cataloguing in Publication Data
Main entry under title:
Flight training manual
4th ed.
ISBN 0-7715-5115-0

1. Airplanes — Piloting. I. Canada. Transport Canada.

TL712.F55 1992   629.132'52   C91-095123-3

Photographs courtesy Aviation Training Systems Ltd.

Catalogue No. T52-14/1994E

1 2 3 4 5 BP 03 02 01 00 99

Printed and bound in Canada

Contents

Preface

The aim of this manual is to provide basic, progressive study material for student pilots preparing for licensing, pilots improving their qualifications, and for the guidance of flight instructors. As such, it complements the Transport Canada Flight Instructor Guide.

This manual provides information and direction in the introduction and performance of flight training manoeuvres as well as basic information on aerodynamics and other subjects related to flight training courses. Thus, a working knowledge of the terms and the material in this manual that are relevant to the training being taken will enable the student to gain maximum benefit from the training.

The contributions by many Canadian flight instructors to the material presented here are gratefully acknowledged.

PART ONE

THE AIRCRAFT AND OPERATIONAL CONSIDERATIONS

CHAPTER ONE

Basic Principles of Flight

The Third Law of Motion

Heavier-than-air flight can be explained by various scientific laws and theorems. Of these, Newton's Third Law of Motion is possibly the fundamental one. "For every action there is an equal and opposite reaction." A propeller accelerates a mass of air backward, and thereby receives an equal forward force. This forward force, called *thrust*, pulls the aircraft ahead.

As the aircraft is thrust forward by the propeller on take-off, the wing meeting the oncoming air begins to generate *lift* (Fig. 1-1). As the forward speed of the aircraft increases, this lift force increases proportionately. When the lift force is equal to the weight of the aircraft, the aircraft begins to fly.

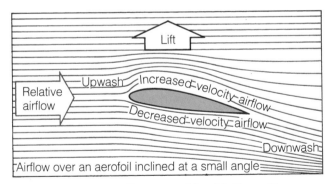

Figure 1-1   Lift Is Generated by Air Travelling Faster above the Aerofoil than below it

The Atmosphere

An aircraft operates in a world that is very near the earth, yet has vastly different properties. This world is the atmosphere, composed of air, which surrounds the earth. We must know something about this atmosphere to understand flight.

Density and Pressure

One property of air that is apt to mislead the novice student of flight is *density*. Ask the weight of the air in an ordinary room and answers will vary from "almost nothing," to "about 10 pounds." Yet the answer is close to 300 pounds and in a large hall may be over a ton! Raise that half a mile above the earth, though, and the air in it will weigh far less. Its density — its mass per unit volume — has changed. These changes in density are measured as *air pressure*.

It is true, of course, that the density of air is low compared with that of water, yet it is this property of air that makes flight possible. Air being the medium in which flight occurs, as its properties change the characteristics of a particular flight will change.

The average pressure at sea level due to the weight of the atmosphere is 14.7 pounds per square inch, a pressure that causes the mercury in a barometer to rise 29.92 inches. In a standard situation the pressure drops from 14.7 pounds per square inch at sea level to 10.2 pounds per square inch at 10,000 feet.

Temperature Changes (Lapse Rate)

With an increase in height there is a decrease in air temperature. The reason is that the sun's heat passes through the atmosphere without appreciably raising the temperature. The earth, however, absorbs the heat. The temperature of the earth is raised and the air in contact with it absorbs some of the heat.

The Four Forces

An aircraft in flight is under the influence of four main forces: *lift, weight, thrust*, and *drag* (Fig. 1-2).

3

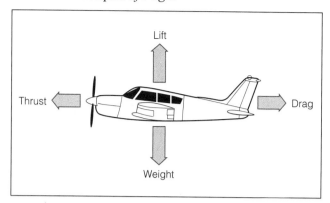

Figure 1-2 The Four Forces

Thrust and Drag Forces

In level flight two principal horizontal forces act on an aircraft — thrust and drag. Thrust is provided by the engine acting through the propeller, and drag by the resistance of the air to the passage through it of the aircraft and all its component parts.

Lift and Weight Forces

The vertical forces acting on an aircraft are lift and weight. Weight is the total weight of the aircraft and its contents; it is considered to act through a single point termed the *Centre of Gravity*. Lift acts at 90 degrees to the *relative airflow* (Fig. 1-3). It is not necessarily perpendicular to the horizon; in flight it may act at a considerable angle to the horizon. For computation purposes, the total force of lift is considered to act through one point of the wing. This point is called the *Centre of Pressure* (Fig. 1-4).

Relative Airflow

Relative airflow is always parallel with and directly opposite to the aircraft's flight path. It has nothing to

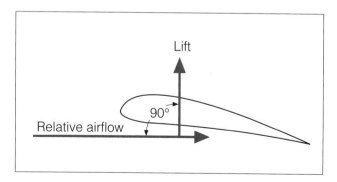

Figure 1-3 Lift Acts at 90 Degrees to the Relative Airflow

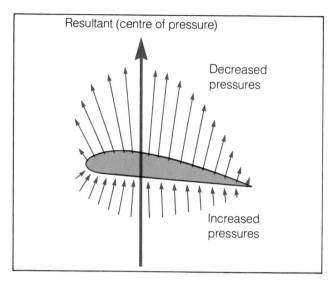

Figure 1-4 Pressure Distribution of an Aerofoil

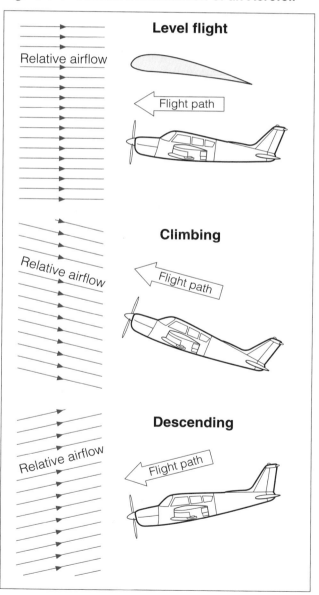

Figure 1-5 Relative Airflow

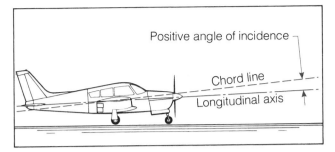

Figure 1-6   Angle of Incidence

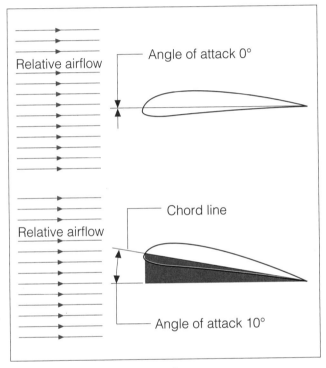

Figure 1-7   Angle of Attack

Lift

Any flat object, such as a flat plate, inclined upward to the relative airflow will provide lift. The kite is an example. The wings of the first aircraft were like a flat plate in this respect. It was later discovered that much more lift could be produced by curving the upper wing surface, and the aerofoil evolved (Fig. 1-8). The curved upper surface also provided for a thicker structure, which allowed for increased strength, fuel storage, and eventually the elimination of exterior structural members.

Simply stated, the wing generates part of the total lift by deflecting air downward. A wing also derives part of its lift from the pressure differential between the upper and lower surfaces. The theoretical expression of this fact is found in *Bernouilli's theorem*. This theorem indicates that as the velocity of air increases, its pressure decreases.

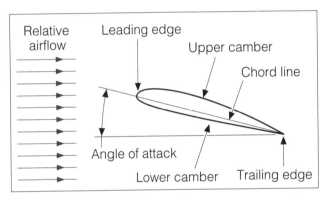

Figure 1-8   Aerofoil Terminology

do with the aircraft's flight attitude, just with the direction of travel of the aircraft (Fig. 1-5).

Angle of Incidence and Angle of Attack

Angle of incidence refers to the fixed angle between the plane of the wing chord and the longitudinal axis of the aircraft (Fig. 1-6). It should not be confused with *angle of attack*, which varies according to the angle between the wing chord and the relative airflow. This angle can vary according to the direction of motion of the aircraft (Fig. 1-7).

Pressure Distribution over an Aerofoil

Due to the curvature of the wing surfaces, the airflow accelerates as it flows around the wing. With the upper surface having a greater camber than the lower surface, the air flowing above the wing will be accelerated more than the air flowing beneath the wing. According to Bernouilli's theorem, this results in a relatively large decrease in pressure above the wing and a smaller decrease in pressure below the wing (Fig. 1-4). This resultant pressure differential produces the force called lift.

Tests also show that as the angle of attack of an aerofoil in flight is increased, the Centre of Pressure moves gradually forward. At a point well beyond the angle of attack for ordinary flight, it begins to move back again. When the Centre of Pressure moves back far enough, the nose of the aircraft will pitch downward suddenly as the wing enters a stalled condition (Fig. 1-9).

The *boundary layer* is a thin layer of air, sometimes no more than a hundredth of an inch (0.25 millimetres) thick, flowing over the surface of a wing in flight. The boundary layer is divided into two parts: (1) the desirable *laminar* layer, and (2) the undesirable *turbulent* layer. Air flowing over the wing begins by conforming to its shape; at this stage the boundary layer is smooth and very thin. This is the laminar layer. There is a point of transition, which moves between the leading and trailing edges of the wing, where the boundary layer

starts to become turbulent and increasingly thick. The airflow beyond this point is described as the turbulent layer.

To maintain a laminar flow over as much of the aerofoil surface as possible, the laminar flow type wing was developed. This design is concerned with the transition point. The laminar flow wing is often thinner than the conventional aerofoil, the leading edge is more pointed, and the section nearly symmetrical, but most important of all, the point of maximum *camber* (the point of greatest convexity of the airfoil from it's chord) is much farther back than on the conventional wing. The pressure distribution on the laminar flow wing is much more even, as the airflow is accelerated very gradually from the leading edge to the point of maximum camber. As the stalling speed of a laminar flow wing is approached, the transition point will move forward much more rapidly than it will on a conventional aerofoil.

Drag

Drag is the force that acts parallel to the relative airflow and retards the forward motion of an aircraft. The total aircraft drag is the sum of *induced* drag and *parasite* drag. Induced drag is a by-product of lift, and parasite drag is made up of all the other drag (Fig. 1-10).

Induced Drag.   Wing tip vortices are formed when higher pressure air beneath the wing flows around the wing tip into lower pressure air above the wings (Fig. 1-11). This disturbed air contributes to induced drag.

In addition to vortices, downwash is produced when air flowing around the wing is deflected downward (Fig. 1-1). Downwash is required in the production of lift and results in induced drag.

A way to visualize induced drag caused by downwash is to picture the resultant lift being tipped backward as the angle of attack is increased (Fig. 1-12). Induced drag is greatest during low airspeed because of the large angle of attack. As speed increases the angle of attack decreases and so does induced drag.

Aspect ratio affects induced drag. The aspect ratio is the ratio of the span to the mean chord (Fig. 1-13). The greater the span of an aerofoil in relation to its chord, the less the induced drag. A long aerofoil with a relatively narrow chord is called a high aspect ratio wing.

Parasite Drag.   This is drag made up of all other drag on the aircraft that is not caused by lift. Unlike induced drag, parasite drag increases as the speed increases.

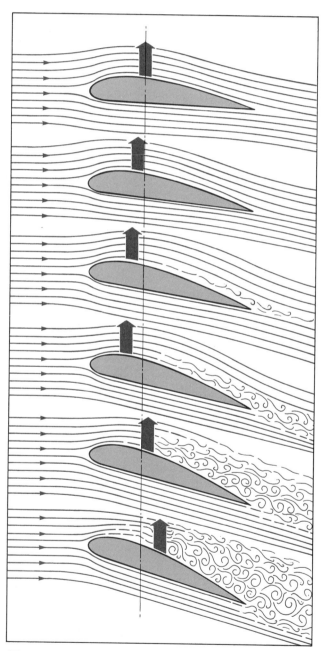

Figure 1-9   The Centre of Pressure shifts as the Angle of Attack Changes

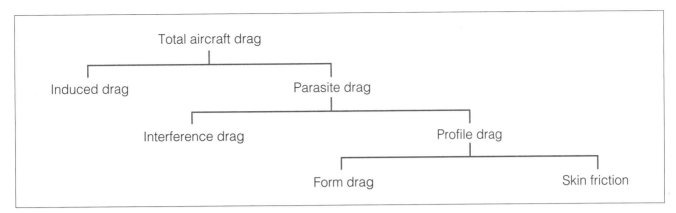

Figure 1-10 Total Drag

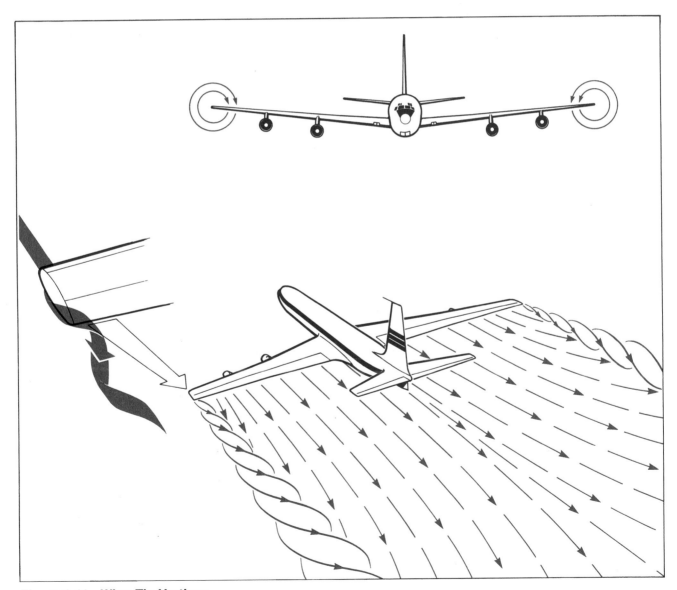

Figure 1-11 Wing Tip Vortices

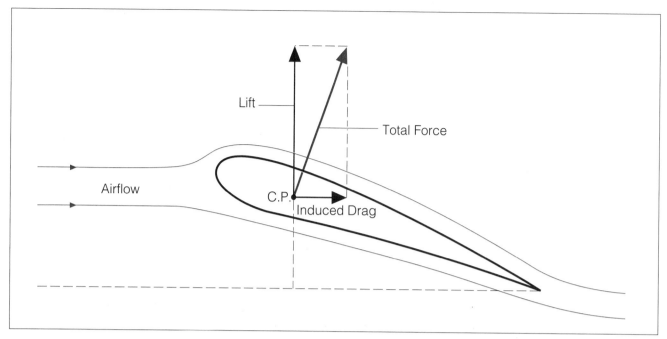

Figure 1-12   Forces Acting on an Aerofoil

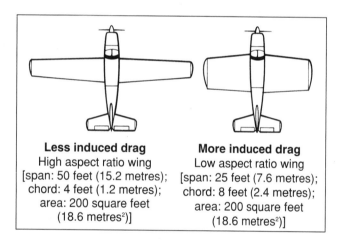

Less induced drag
High aspect ratio wing
[span: 50 feet (15.2 metres);
chord: 4 feet (1.2 metres);
area: 200 square feet
(18.6 metres²)]

More induced drag
Low aspect ratio wing
[span: 25 feet (7.6 metres);
chord: 8 feet (2.4 metres);
area: 200 square feet
(18.6 metres²)]

Figure 1-13   Aspect Ratio

Interference Drag.   Interference drag is a result of the interference of airflow between two sections of the aircraft. For example, where the wing and fuselage come together air flowing along the fuselage will interfere with the air flowing over the wing.

Profile Drag.   Profile drag consists of *form drag* and *skin friction*.

Form Drag.   Form Drag is caused by the form or shape of a body as it resists motion through the air. Streamlining of all parts of the aeroplane that are exposed to the air will greatly reduce this type of drag.

Skin Friction.   Skin friction is the tendency of air to hold an aircraft back by clinging to its surfaces. A smooth and highly polished aircraft will be affected

much less by this type of drag. Ice, dirt, or even insects accumulated on an aircraft contribute to skin friction.

Minimum Drag.   Since induced drag decreases with an increase in airspeed and parasite drag increases with an increase in airspeed, there is an airspeed where the total drag is the lowest (Fig. 1-14). This minimum drag speed will be discussed further in the exercises on gliding, range and endurance.

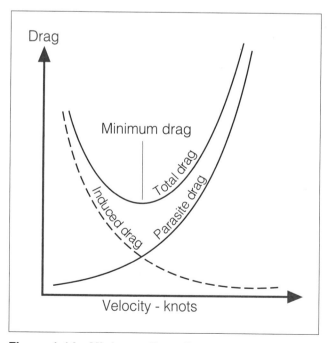

Figure 1-14   Minimum Drag Speed

Equilibrium

A body at rest tends to remain at rest and a body in motion tends to remain in motion in a straight line unless an external force is applied. This is Newton's First Law of Motion. A body that is neither accelerating nor decelerating may be said to be in equilibrium. A parked aircraft is in equilibrium; an aircraft in straight-and-level flight at a constant airspeed is in equilibrium; an aircraft in a straight descent or climb at a constant airspeed is also in equilibrium. However, an aircraft in a turn at a constant height and airspeed is not in equilibrium, since during a co-ordinated turn the aircraft is always accelerating toward the centre of the turn.

A lot of effort is invested in designing aircraft to maintain a state of equilibrium. However, a pilot must be able to disturb this equilibrium to manoeuvre the aircraft. Flight controls allow a pilot to produce forces about the three axes of the aircraft (Fig. 1-15). These forces disturb the aircraft's equilibrium and allow the aircraft to be manoeuvred.

Roll, Ailerons, Longitudinal Axis

When an aircraft is *rolled*, one aileron is depressed and the opposite one is raised. The "down" aileron increases the effective camber of its wing producing more lift than the other wing. The "up" aileron reduces camber producing less lift. As a result, the "down aileron" wing rises and the "up aileron" wing moves down. The total effect causes the aircraft to roll about its longitudinal axis. Aileron movement is controlled by rotation of the control wheel or left or right movement of the control column. When the control wheel is rotated to the left, or the control column moved to the left, the left aileron rises and the aircraft rolls to the left.

Pitching, Elevators, Lateral Axis

Backward movement of the control column or wheel raises the elevators. This changes the camber, producing a force that causes the tail to go down and the nose to rise. Forward movement of the control lowers the elevators; this produces the opposite reaction, raising the tail and lowering the nose. The elevators produce and control a pitching movement about the lateral axis of the aircraft.

Yaw, Rudder, Normal (Vertical) Axis

The left or right movement of the nose of an aircraft in flight is controlled by the rudder, through the rudder pedals. The rudder is hinged to the trailing edge of the fin (vertical stabilizer). Foot pressure on the left rudder pedal causes the rudder to move to the left and introduce camber to the fin; this causes a mass of air to be accelerated to the left which (Newton's Third Law) moves the tail of the aircraft to the right and causes the nose to yaw (move) to the left. Opposite reactions occur when pressure is applied to the right rudder pedal.

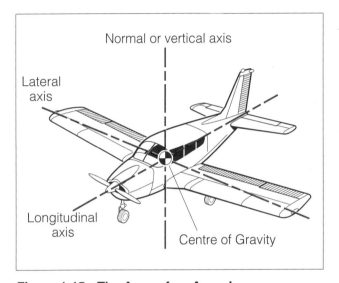

Figure 1-15 The Axes of an Aeroplane

Adverse Yaw

Any yaw, regardless of origin, having an effect contrary to the interests of the pilot is called *adverse yaw*. When adverse yaw occurs, the rudder is used to counteract its effect and help provide directional control. Adverse yaw results from *aileron drag, gyroscopic precession, torque, slipstream*, and *asymmetric thrust*.

Aileron Drag. Flight controls are designed to be effective, well balanced, and responsive; however, the ailerons have an operating characteristic that should be explained. For example, if a turn to the left is desired, movement of the control column to the left causes the right aileron to move downward and increase the camber of the right wing, causing that wing to develop more lift and rise up (Fig. 1-16). Conversely, the left aileron moves upward and decreases the camber of the left wing, causing that wing to develop less lift and move down. However, in developing more lift, the right wing is subjected to more induced drag, and in developing less lift, the left wing is subjected to less induced drag. The whole effect causes a momentary yaw to the right, when a turn to the left is desired.

Yaw caused by aileron drag may be more noticeable when operating at reduced airspeeds, or when applying large, abrupt aileron deflection. When aileron drag causes the aircraft to yaw, rudder pressure must be applied simultaneously with, and in the same direction as, aileron input. The rudder pressure required will vary and is necessary only while aileron control is being applied. When aileron pressure is removed aileron drag ceases, and the rudder pressure applied to correct aileron drag must be readjusted to maintain co-ordinated flight.

In most aircraft of recent manufacture, using one or a combination of two design features partly compensates for aileron drag. Differential ailerons are designed to cause the downgoing aileron to move through a smaller angle than the upgoing aileron for a given movement of the control column. The upgoing aileron produces more drag and helps to minimize adverse yaw. Frise ailerons produce a similar effect by placing the hinge such that the nose of the upgoing aileron projects into the airflow beneath the wing and produces extra drag, while the downgoing aileron is streamlined.

Gyroscopic Precession.   When a force is applied to a spinning gyro wheel, it will react as though the force

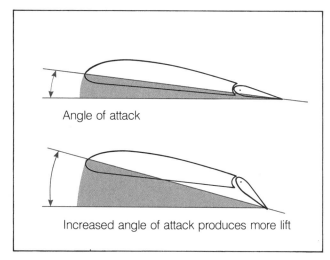

Angle of attack

Increased angle of attack produces more lift

Figure 1-16   The Effect of the Aileron on the Angle of Attack

had been applied in the same direction at a point 90 degrees from where the force was actually applied. This is called gyroscopic precession. The spinning mass of an aircraft propeller is an example of a "gyro wheel." As such, it is susceptible to gyroscopic precession. Gyroscopic effect of this kind can sometimes be quite noticeable in tail wheel equipped aircraft as the tail is raised during the take-off sequence (Fig. 1-17). It is as though the pilot had reached out and applied the force to the propeller at the top of the arc. If the propeller rotates clockwise, when viewed from the

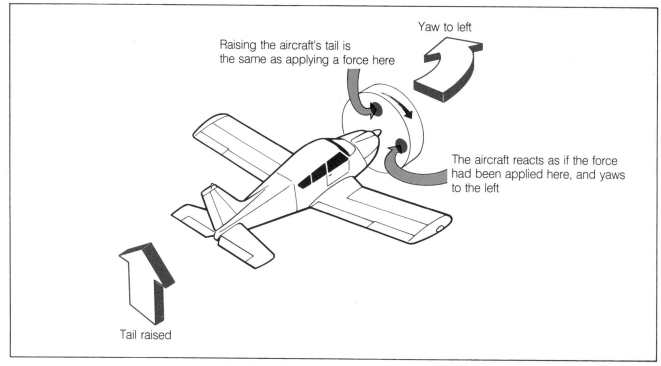

Yaw to left

Raising the aircraft's tail is the same as applying a force here

The aircraft reacts as if the force had been applied here, and yaws to the left

Tail raised

Figure 1-17   The Effect of Raising the Aircraft's Tail

cockpit, gyroscopic effect changes the position of the force and applies it to the extreme right point of the propeller arc. This causes the aircraft to attempt to turn (yaw) to the left and requires application of right rudder pressure to maintain the desired heading.

Torque.   The engine rotates the propeller in one direction, but in so doing, and in obedience to Newton's Third Law, it also tries to rotate the whole aircraft in the opposite direction. In the case of most North American aircraft, if the propeller rotates clockwise, a downward force is being exerted on the left side of the aircraft.

Under conditions of high power while the aircraft is on the ground, torque will cause the left wheel to carry slightly more weight than the right. This will produce more friction or drag on the left wheel and add to the tendency of an aircraft to yaw to the left on the take-off roll.

In the design of the aircraft this undesirable force is sometimes neutralized by giving the left wing slightly more angle of incidence and, therefore, slightly more lift than the right wing. It should be noted that torque does not directly cause yaw. Torque causes roll, and roll in turn causes yaw.

Slipstream.   The mass of air thrust backward by the propeller is called the slipstream. It is roughly the size of a cylinder of the same diameter as the propeller.

The velocity of the slipstream is greater than that at which the aircraft is travelling through the air. This means that the velocity of the air flowing over those parts of the aircraft in the slipstream would be much more than that of the airflow over parts not in the slipstream.

The propeller imparts a rotary motion to the slipstream in the same direction as the propeller is turning. The result is that the slipstream strikes only one side of aircraft surfaces, such as the fin and rudder, and affects the directional and lateral balance of the aircraft (Fig. 1-18). To compensate for this the fin or engine may be offset slightly to balance the aircraft for normal cruising flight. This balance is upset when the engine power is changed above or below cruise power settings (Fig. 1-19).

Asymmetric Thrust.   Adverse yaw is also caused by the asymmetrical loading of the propeller. When an aeroplane is flying at a high angle of attack (with the propeller axis inclined) and with high power settings, the downward moving blade, which is on the right side of the propeller as seen from the cockpit, has a higher angle of attack and therefore produces more thrust than the upward moving blade on the left. The result is a yawing tendency of the aircraft to the left.

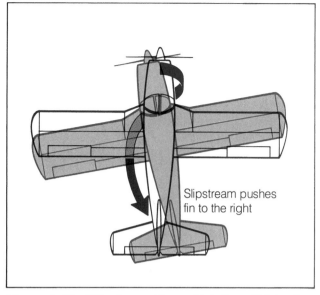

Slipstream pushes fin to the right

Figure 1-18   Slipstream Has a Rotational Velocity

Brakes

On many aircraft each main landing wheel has its own independent braking system to facilitate manoeuvring on the ground. Brakes of this type may be used to shorten a landing roll and give directional control on the ground at speeds where rudder control is inadequate. Pressure applied to the left brake pedal brakes the left wheel and turns the aircraft to the left; pressure applied to the right brake pedal turns the aircraft to the right. To bring an aircraft to a straight stop, equal or near equal pressure must be applied to each brake pedal.

Trim Tabs

To improve control and balance *(trim)* of an aircraft, small auxiliary control surfaces called *trim tabs* are fixed or hinged to the trailing edges of the ailerons,

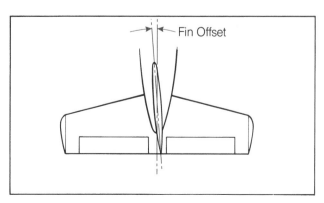

Fin Offset

Figure 1-19   Fin Offset

elevators, and rudder. Fixed tabs are pre-set on the ground to obtain a balanced control loading at the normal level cruising speed of particular aircraft. Hinged tabs are controlled by the pilot. Larger aircraft, for the most part, have hinged tabs fitted to all control surfaces to compensate for lateral shifts in loading and to provide additional rudder control in the event of an engine failure on multi-engined aircraft. In the case of most small single-engined aircraft, such as those used for flight training, only the elevators have controllable trim. Elevator trim compensates for the constantly changing longitudinal stability resulting from varying attitudes of flight. Fixed trim tabs, if fitted, are normally adequate for the lateral (aileron) and directional (rudder) stability and control of this class of aircraft.

Variable Incidence Tail Planes (Horizontal Stabilizer)

On some aircraft the incidence of the tail plane can be varied in flight to trim the aircraft longitudinally. The effect is much the same as trimming the elevators on an aircraft with a fixed tail plane.

Flaps

Flaps are controlled by the pilot. They improve the lift and other characteristics of an aerofoil by increasing the camber of a large portion of the wing. Some of the operational advantages of flaps are:

1. Stalling speed is decreased.
2. A steeper approach to landing can be made without an increase in airspeed.
3. Forward visibility is improved on approach to landing due to the lower position of the nose.
4. The take-off run may be shortened.

The plain flap is actually a portion of the main aerofoil, including upper and lower surfaces, which hinge downward into the relative airflow. However, in the case of more sophisticated articulated flap systems (Zap and Fowler are typical examples), there is an effective increase of the chord of the aerofoil, which up to a given point in their operation greatly increases lift while imposing minimum drag (Fig. 1-20).

When the pilot selects a flap position, both flaps go down or up together. When flaps are fully retracted (up) they conform to the shape of the wing. Flaps must be used judiciously at all times but extreme care must be taken when retracting them in flight — especially near the ground — because of the sudden loss of lift and change in the aircraft's balance.

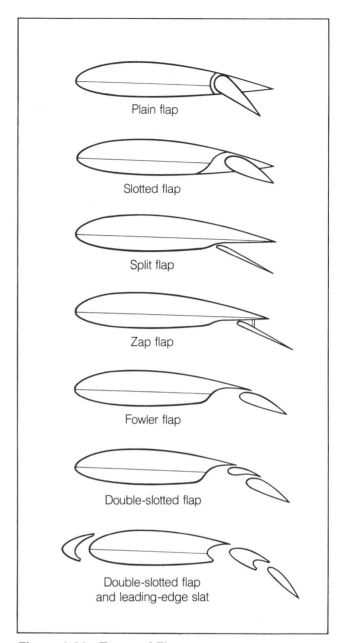

Figure 1-20   Types of Flaps

Stability

A *stable* aircraft is one that tends to return to its original flight condition after being displaced by some outside force, such as an updraft or other air disturbance. The stability of an aircraft concerns its three planes of rotation: (1) pitching, (2) yawing, and (3) rolling. Stability in the pitching plane is called *longitudinal* stability, in the rolling plane, *lateral* stability, and in the yawing plane, *directional* stability.

Lateral Stability.   One design feature that provides stability in the rolling plane is *dihedral*. This is a wing design feature in which the wing tips are higher than the centre section of the wing. When a wing is lowered due to an outside influence, such as turbulent air, the aircraft starts slipping toward the "down" wing; the "down" wing is now exposed to more airflow (than the "up" wing), and, as a result of its higher angle of attack, produces more lift and returns to its former level state (Fig. 1-21).

Longitudinal Stability.   Of all the characteristics that affect the balance and controllability of an aircraft, longitudinal stability is the most important, as it can be influenced by both aerodynamic and physical factors, including human error.

The position of the Centre of Gravity of an aircraft has the greatest influence on its longitudinal stability, but this stability is also influenced by changes of speed, power, and attitude. It is difficult to obtain the right degree of longitudinal stability to meet all conditions of flight, but it is essential to achieve an acceptable compromise if the aircraft is to be safe and pleasant to handle. Turbulent air, operation of the flaps, and other factors can disturb the balance of the aircraft.

This problem is primarily resolved by the horizontal stabilizer (tail plane) aided by the vertical stabilizer (fin). Purposely placed at a considerable distance behind the wing, these stabilizers aerodynamically provide the basic forces necessary to counteract the effect of outside forces. Because of its distance from the Centre of Gravity, which gives it great leverage, even a small force on the tail plane will produce a large correcting moment.

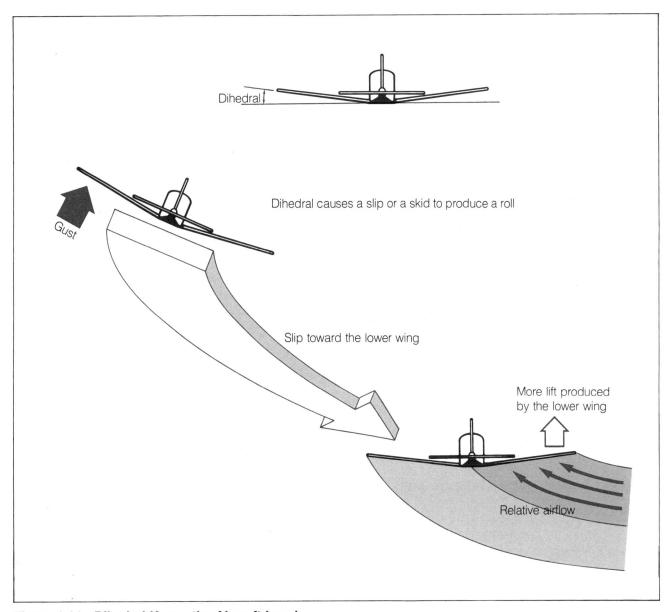

Figure 1-21   Dihedral Keeps the Aircraft Level

Much like the tail feathers on an arrow, the tail plane will resist outside influences altering the aircraft's selected longitudinal flight path. Outside influences and forces may be likened to unco-ordinated use of the flight controls, to which a stable properly trimmed aircraft will also offer resistance. A stable aircraft will not attempt to counteract forces intentionally introduced by co-ordinated use of the flight controls and throttle.

Directional Stability.   Directional stability concerns the motion of the aircraft about the normal axis, or the yawing motion of the aircraft. An automobile has a directional stability that can be seen every time the car turns a corner. After the turn is made and the steering wheel released, the wheels straighten and the car moves in a straight direction. This is directional stability. The vertical stabilizer contributes to the directional stability of an aircraft.

Load Factors

Any force applied to an aircraft to deflect its flight from a straight line produces stress on its structure, the amount of which is termed a *load factor*. Load factor is the ratio of the load supported by the aeroplane's wings to the actual weight of the aircraft and its contents.

A load factor of 3 means that the total load on an aircraft's structure is 3 times its gross weight. Load factors are generally expressed in terms of "G". A load factor of 3 is usually spoken of as 3 Gs. When an aircraft is subjected to 3 Gs, for example, in an abrupt pull-up, the pilot will be pressed down into the seat with a force equal to 3 times his or her own weight. Thus, a rough estimate of the load factor obtained in a manoeuvre can be made by considering the degree to which a person is pressed down in the seat.

Load factors are important for two distinct reasons: (1) a structural overload can be imposed upon an aircraft; (2) load factor increases the stall speed making stalls possible at seemingly safe airspeeds.

Fig. 1-22 shows an aircraft banked at 60 degrees. The total lift produced is 4,000 pounds (1814 kilograms). The aircraft weight is 2,000 pounds (907 kilograms). The ratio of total lift to weight is **2**. So we say the load factor is **2**.

Fig. 1-23 reveals an important fact about turns — the load factor increases at a tremendous rate after the bank has reached 50 degrees. It is important to remember that the wing must produce lift equal to the load factor, otherwise it will be impossible to maintain altitude. Notice how rapidly the line representing load

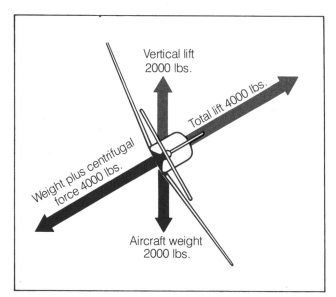

Figure 1-22   Load Factor in 60 Degree Banked Turn

factor rises as it approaches the 90 degree bank line, which it reaches only at infinity.

Therefore, although an aircraft may be banked to 90 degrees, a constant altitude turn with this amount of bank is mathematically impossible for conventional aircraft. At slightly more than 80 degrees of bank the load factor exceeds 6 Gs, which is generally the flight load factor limit of aircraft structurally designed for aerobatic flight. For conventional light aircraft, the approximate maximum bank, in a sustained level co-ordinated turn, is 60 degrees. An additional 10 degrees of bank will increase the load factor by approximately 1 G, bringing the loading close to the point at which structural damage may occur.

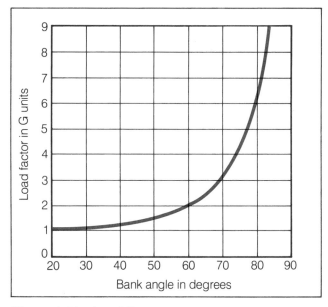

Figure 1-23   Load Factors Produced at Varying Degrees of Bank at Constant Altitude

Stalling Speeds

When the angle of attack of the wing exceeds the stalling angle, the relatively smooth flow of air over the wing breaks up and tears away, producing the sudden change of flight characteristics and loss of lift known as a *stall*. Within the limits of its structure and the physical strength of the pilot, any aircraft may be stalled at any airspeed.

The stalling speed of an aircraft increases in proportion to the load factor. An aircraft with a normal stalling speed of 50 KT can be stalled at 100 KT by imposing a load factor of 4 Gs upon it. If it were possible to impose a load factor of 9 Gs upon this aircraft, it could be stalled at 150 KT. This knowledge must be applied from two points of view: (1) the danger of unintentionally stalling an aircraft by increasing the load factor, as in a steep turn or a spiral; (2) the tremendous load factor imposed upon an aircraft when it is intentionally stalled above its manoeuvring speed. This may be done by an abrupt pull-up, or any other manoeuvre producing a load factor beyond 1 G, and can result in a hazardous and sudden loss of control.

Abrupt or excessive deflection of the flight controls can impose severe structural loads upon an aircraft, and these load factors are directly proportionate to the aircraft's speed. This type of loading may occur when executing certain flight manoeuvres or during flight in turbulent air. For this reason each aircraft type has a design manoeuvring speed, which is defined as "the maximum speed at which the flight controls may be fully deflected without causing structural damage." This speed limitation may be found in the Aircraft Flight Manual of a particular aircraft.

Since leverage in control systems varies from aircraft to aircraft, the pressure required on controls cannot be accepted as an index of the load factors produced by the various manoeuvres of any particular aircraft. Load factors can be measured by certain instruments, but since these instruments are not common in conventional training aircraft, it is important to develop the ability to judge load factors from their effect on the body.

Turbulence

Aircraft are designed to take *gust loads* of considerable intensity. Gust loads represent loading imposed upon an aircraft, particularly the wings, as a result of the aircraft being flown into vertical or horizontal air currents, commonly referred to as "turbulence." By definition, gust load factor is an acceleration imposed upon an aircraft flown into a gust. Gust load factors increase as airspeed increases: in moderate or extreme turbulence, such as may be encountered near thunderstorms or frontal weather conditions, it is wise to reduce airspeed to the manoeuvring speed specified for the aircraft. This is the speed least likely to permit structural damage to the aircraft, yet it allows a sufficient margin of safety above the normal stalling speed if abrupt control movements must be made. "Maximum dive" or "never exceed" speeds for a particular aircraft are determined for smooth air only. Abrupt manoeuvring or high diving speeds in turbulent air at airspeeds above the specified manoeuvring speed can place damaging stresses on the whole structure of an aircraft.

Structural Damage

Many people have a mistaken tendency to consider load factors only in terms of their immediate effect upon aircraft components. The cumulative effect of exceeding the load factor over a long period of time may loosen and weaken vital parts so that a structural failure may occur later, when the aircraft is being operated normally.

CHAPTER TWO

Weight and Balance

Centre of Gravity

It is the responsibility of the pilot-in-command to load the aeroplane in accordance with the weight and balance information provided by the manufacturer. This information limits the maximum load that can be carried. It also limits the placement of the load within the aeroplane. In order to do this, the pilot-in-command must ensure that the weight is below the gross weight of the aeroplane and the Centre of Gravity is within its specified range.

To simplify loading problems, most manufacturers of light aircraft supply pre-calculated graphs, charts, or loading examples for specific aircraft that are adequate for the use of the average private pilot. However, you should have a working knowledge of the basic principles behind these calculations.

The Centre of Gravity for each aircraft is calculated at the factory and recorded on the weight and balance report. The method used to find the Centre of Gravity for an empty aircraft is also used to calculate the shift of the Centre of Gravity caused by loading of fuel, pilot, passengers, baggage, etc.

It is imperative, for optimum control response and stability, that the Centre of Gravity of an aircraft be maintained within its permissible design limitations. You can seriously affect the controllability of your aircraft by positioning the load incorrectly. If the Centre of Gravity loading limitations as outlined in the Aircraft Flight Manual are followed carefully, you will have a stable aircraft and predictable response to the controls. If the Centre of Gravity is permitted to go beyond the forward limitations, the aircraft will become less responsive to up elevator control, and very difficult to trim.

However, in the average light aircraft there is considerably more scope for error in exceeding the *aft* Centre of Gravity limitations. Careless aft loading can lead to very hazardous balance and control problems, even though the maximum permissible overall weight is not exceeded. For example, if the fuel load of a four place aircraft is reduced to remain within maximum permissible take-off weight, then two relatively lightweight people are placed in the forward seats, two very heavy people are placed in the rear seats, and the aft luggage compartment is loaded to its maximum allowable weight. An aircraft so loaded could have such an excessive aft Centre of Gravity that the pilot would have control problems beginning from take-off, and the normal stall characteristics might change drastically. Should an aircraft with excessive aft loading be permitted to enter a spin, there is every possibility that recovery would be extremely difficult if not impossible to execute.

Useful Load

Of prime interest to the pilot-in-command is the *useful load* an aircraft can carry. In light aircraft, useful load consists of crew, passengers, baggage, usable fuel and oil, and other non-fixed items. It may be calculated by subtracting empty weight from maximum permissible weight. These weights are generally defined as:

1. Licensed Empty Weight.   The weight of the basic aircraft, including its fixed equipment and unusable fuel and oil.

2. Basic Empty Weight.   The weight of the basic aircraft, including its fixed equipment, unusable fuel and full operating fluids, including full oil.

3. Maximum Permissible Weight.   The maximum permissible gross take-off weight specified in the aircraft's Certificate of Airworthiness.

Passengers.   Although airlines use average passenger weights for weight and balance computations, actual passenger weights must be used for aircraft with limited seating capacity. Light aircraft can easily be loaded outside limits when estimates or average weights are used, particularly when winter clothing is worn. If in doubt, ask passengers how much they weigh and make appropriate allowances for clothing.

Weight and Balance

The maximum permissible weight must never be exceeded. If you carry additional fuel to give the aircraft more range, you must in many cases balance it by reducing the number of passengers or baggage, freight, or other such weight, so as not to exceed the maximum permissible weight.

The pilot-in-command must know all the loading information about the aircraft to be flown and be able to determine permissible loading and its correct disposition. Current weight and balance data should be carried as part of the aircraft documentation; unless loaded in accordance with this information, the aircraft cannot be considered airworthy. In addition, it may bear a placard concerning operational loading, such as the seat to be occupied in solo flight or a fuel tank to be emptied first. The information on these placards must be observed scrupulously.

Weight and balance limitations are imposed for the following principal reasons:

1. The effect of the disposition of weight (and subsequently balance) on the flight characteristics of the aircraft, particularly on stall and spin recoveries, slow flight, and stability.
2. The effect of the weight on primary and secondary structures of the aircraft.
3. The effect of weight on take-off and landing performance.

Computing Weight and Balance

The weight and balance calculations for individual flights are computed by using information in the aircraft's Weight and Balance Report, and the weights and disposition within the aircraft of each passenger and/or item of the load proposed for the flight. The items of information available on the Weight and Balance Report are:

1. Empty weight, in pounds.
2. Balance Datum. This is the reference point from which all weight and balance calculations are made. It could be anywhere on the aircraft, preferably somewhere forward of the Centre of Gravity. It could even be a point in open space several feet

in front of the nose. To avoid negative numbers, most aircraft balance datums are situated so that all useful loads are positioned aft of the balance datum.
3. Centre of Gravity (empty), in inches from the balance datum.
4. Moment arm. The distance in inches from the balance datum to the Centre of Gravity of the aircraft or Centre of Gravity of an item of load.

Concerning the proposed load, the pilot-in-command must be aware of:

1. The weight in pounds of each passenger and/or item.
2. The location of each passenger and/or item within the aircraft.

Weight and balance factors may be computed as follows:

1. The balance moment of the empty aircraft is found by multiplying the empty weight by the moment arm of the aircraft.
2. The balance moment of each item of load is found by multiplying its weight by its respective moment arm.
3. The new Centre of Gravity is found by dividing the total balance moment by the total weight of the aircraft.

Sample Computations. A sample weight and balance calculation for a typical light aircraft, with a Centre of Gravity 30 inches (0.762 metres) aft of its datum when empty, using the foregoing items and factors, might look like this:

	Weight (LB)	Moment Arm (inches)	Balance Moment (LB-inches)
Aircraft empty	1,000	30	30,000
Pilot	170	20	3,400
Passenger	190	20	3,800
Fuel	50	30	1,500
Oil	10	20	200
	1,420		38,900

Metric Version

	Weight (kg)	Moment Arm (metres)	Balance Moment (kg-metres)
Aircraft empty	453.6	0.762	345.6
Pilot	77.1	0.508	39.0
Passenger	86.2	0.508	43.6
Fuel	22.7	0.762	17.3
Oil	4.5	0.508	2.3
	644.1		447.8

The new Centre of Gravity is 27.4 inches (0.695 m) aft of the balance datum. Divide 38,900 (447.8) by 1,420 (644.1).

The pilot-in-command must now refer to the current aircraft documentation and verify that the loaded weight of 1,420 pounds (644.1 kilograms) and the loaded Centre of Gravity of 27.4 inches (0.695 metres) aft of the datum are both within the prescribed tolerances. If they are not, the aircraft should not be flown until satisfactory adjustments are made.

The following is an example of another weight and balance computation for a more sophisticated aircraft:

	Weight (LB)	Moment Arm (inches)	Balance Moment (LB-inches)
Aircraft empty	1,552	37	57,424
Pilot and passenger at 185 LB each	370	36	13,320
One passenger in rear seat	140	70	9,800
Baggage	120	95	11,400
Fuel, 55 US gals. at 6 LB/gal.	330	48	15,840
Oil, 2 US gals. at 7.5 LB/gal.	15	−15	−225
	2,527		107,559

Metric Version

	Weight (kg)	Moment Arm (metres)	Balance Moment (kg-metres)
Aircraft empty	704.0	0.940	661.8
Pilot and passenger at 84 kg each	168.0	0.914	153.5
One passenger in rear seat	63.6	1.778	113.0
Baggage	54.5	2.413	131.5
Fuel, 208 L at 0.72 kg/L	150.0	1.219	183.0
Oil, 7.6 L at 0.95 kg/L	7.2	−0.381	−2.7
	1147.3		1240.1

In this case the loaded Centre of Gravity (loaded moment divided by loaded weight) is 42.5 inches (1.08 metres) aft of the datum point. The example above uses a balance datum somewhere aft of the nose to illustrate a minus item; in this aircraft the oil reservoir is forward of the balance datum and therefore must be shown as a minus quantity on the total scale of balance.

The pilot-in-command of this particular aircraft must now refer to the current aircraft documentation and verify that the loaded weight of 2,527 pounds (1147.3 kilograms) and the loaded Centre of Gravity of 42.5 inches (1.08 metres) are both within the prescribed tolerances. If they are not, the aircraft should not be considered airworthy until satisfactory adjustments are made.

CHAPTER THREE

Engine Handling

A typical four-cylinder aircraft engine has over 250 moving parts and 70 non-moving parts. The failure of any part may result in a complete loss of power, or sufficient power loss to require an immediate landing. However, because of compulsory testing of material and parts, a high degree of quality control is achieved, resulting in the aircraft engine being one of the most reliable mechanical components in use today. Whether or not this high level of reliability is sustained depends to a great extent on the pilot-in-command's handling of the engine. Besides its flight operation, the handling of an engine includes the use of recommended fuels and oils, pre-flight inspections, and a basic knowledge of how an engine and its ancillary components work. Since it represents the majority of training aircraft power plants in use today, the engine we will consider is air cooled, has horizontally opposed cylinders, and is unsupercharged. The propeller may be made of wood, composition material, or metal and has a fixed pitch.

Inspection

Before any flight the pertinent log-books are studied to check the engine hours, what inspections, repairs, or modifications have been made, whether any reported defects are outstanding, and whether the aircraft has been currently signed out by the appropriate authority.

Never inspect an aircraft engine and propeller until the ignition (magneto) switches have been checked and are off, and even then always treat the propeller as if it were "live." There is not a great deal that a pilot can do in the way of mechanically inspecting an engine prior to flight, but the few things that can be done are extremely important. The engine oil can be checked for acceptable level and the carburettor air filter checked for obstructions. Drain a small sample of fuel from the fuel strainer drain into a suitable glass container. Make sure that the fuel is free of water and sediment. Look for oil and fuel leaks. Physically check components on the engine to see if there are any loose items or wires, etc. Check the propeller and spinner for nicks and security. If in doubt concerning damage to any part of the aircraft, check with your flight instructor or an Aircraft Maintenance Engineer.

Prior to starting an engine there are several things that can be done to ensure that it will perform properly:

1. Verify the fuel supply by a physical check of the fuel tanks.
2. Fuel tank air vents must be open and clear of foreign material to ensure that fuel may flow at recommended rates.
3. Check all fuel tank caps for security. In most cases, if a cap comes off in flight, the contents of the tank will empty rapidly through the filler neck, due to the syphoning action of the airflow.
4. Check the propeller for nicks and other damage that may cause imbalance and undesirable and often dangerous engine vibration.
5. Engine oil. Never add a detergent oil to an engine that uses a non-detergent oil as its regular lubricant. Add non-detergent oil to an engine that uses detergent oil as its regular lubricant in an emergency only.
6. Cold oil. The oil used in the engine of an aircraft is of higher viscosity than that used in most other engines and becomes very thick when cold. With the ignition switch off and mixture control in idle cut-off position, turn the propeller by hand for several revolutions to help break the drag created by cold oil between the piston and the cylinder wall. This will ease starting and reduce the load on the starting mechanism and battery. Remember to treat the propeller as if it were "live."
7. Drum fuelling. Refuelling an aircraft from drums is not a preferred method if regular fuelling

facilities are available, because condensation and flakes of rust are often present in the drums. Since most of the foreign material settles to the bottom of the drum, make sure that the suction tube on the pump being used has at least an inch clearance from the bottom of the drum. A chamois strainer should be used, since it not only removes solids but also resists the passage of water. When refuelling from drums it is critical that the drum, pump, hose, nozzle, and aircraft are properly grounded and bonded. The proper sequence to be followed is: drum to the ground (anchor post), drum to pump, pump to aircraft, nozzle to aircraft, then open the aircraft fuel cap. When finished, reverse the order. Information regarding aviation fuel handling is contained in the *Aeronautical Information Publication* (A.I.P.) Canada AIR section.

8. Ensure that the fuel is of the octane rating specified for the engine. Never use a lower grade; in an emergency use the next higher grade. The octane rating of a fuel is indicated by its colour.

Starting

The engine is started (and operated) as specified in the Aircraft Flight Manual or as specified by a particular operator to meet the requirements of non-standard conditions, such as temperature and elevation extremes.

Once started, use the RPM recommended in the Aircraft Flight Manual for engine warm-up. Too low an RPM results in inadequate distribution of the sluggish engine oil; too high an RPM can cause excessive wear of parts that depend on gradually acquiring heat to expand to operating clearances. If, after starting, an oil pressure indication as specified in the Aircraft Flight Manual is not evident, shut down the engine.

To ensure that there are no fuel blockages between the fuel tanks and the engine, you may wish to run the engine for a period of time on each tank separately while on the ground. In this case you should ensure that the fuel selector is moved to the proper setting before the run-up check and take-off.

Avoid starting and running up an engine where the propeller may pick up loose stones, blow them back, and possibly damage the aircraft or other property behind it.

Aircraft Engine Primers

A fuel system component of many light aircraft is a small hand pump called a *primer*. It is generally located on the instrument panel. The primer draws filtered fuel from the fuel system and injects a fine spray directly into the engine intake ports. This system is useful particularly for cold weather starts when fuel is difficult to vaporize.

Some points to remember regarding the use of primers are:

1. The primer must be used only as specified in the Aircraft Flight Manual.
2. Overpriming will increase the possibility of an engine fire during start.
3. Most manual primers are equipped with a lock and after being pushed full in must be rotated either left or right until the pin is past the notch and the knob cannot be pulled out. If the primer is not locked the engine may draw fuel through the priming system, and the enriched fuel/air mixture can result in engine roughness or engine failure.

Warm-up

There are two warm-up phases. The first is before taxiing to ensure that the engine instruments are beginning to register somewhere in the operating range before applying power to begin taxiing. This warm-up is important when more power may be required to pull the aircraft through snow or loose soil.

The second warm-up, prior to take-off, ensures that engine temperatures and pressures are within the specified limits. At this point the RPM may be increased to hasten the warm-up. The ground running of the engine should be carried out with the aircraft headed into wind or as close to it as possible. As an aircraft engine is closely cowled for efficient in-flight cooling, take care to avoid overheating it on the ground.

Run-up

If there is to be a change in fuel tanks before take-off, change them before the run-up. Should a fuel system malfunction occur as a result of changing tanks, let this show itself during the run-up, not during the take-off.

Carburettor Heat Check

Set the engine RPM as recommended in the Aircraft Flight Manual; if there is no recommended RPM available use the setting recommended for the magneto check. Then:

1. Select "full cold" position of carburettor heat and note RPM.
2. Select "full hot" position and note decrease in RPM; allow RPM to stabilize in this position.
3. Select "cold" position again and note increase in RPM to confirm that the unit and its controls are functioning through their full range. If, on returning the control to "cold," the RPM shows an increase over the initial RPM reading, carburettor icing conditions exist and additional care will be necessary.

The engine air intake filter is usually bypassed when "hot" is selected. Unless otherwise recommended in the Aircraft Flight Manual, use the "cold" position while taxiing or during sustained ground operation of the engine. This will help prevent particles of foreign material, such as sand, from entering the carburettor and engine.

Application of carburettor heat will result in an RPM decrease. If there is no decrease in RPM, suspect a malfunction. Should you suspect that an in-flight power loss is due to an engine air intake filter clogged with snow or ice, apply full carburettor heat to obtain an alternate source of intake air. Carburettor heat should be applied at any time when a power loss due to ice is either noted or suspected. Depending on atmospheric conditions, many pilots perform periodic carburettor heat checks for ice accumulation every ten to fifteen minutes. Under certain conditions, it may be necessary to perform this check more often or even fly with the carburettor heat control in the full hot position.

Magneto Check

The primary purpose of dual ignition in aircraft engines is safety, and the magneto check tests this feature. The magnetos are both operating when the magneto switch is selected to the "both" position. When the magneto switch is selected to "left" or "right," the engine is operating on one magneto only. By selecting one or the other, you can test the proper functioning of each. The first check should be made at low RPM (idle or slightly above) before leaving the flight line to ensure both are working. Next, check the magnetos during the run-up as recommended in the Aircraft Flight Manual. Normally this involves a check of each ignition system for smooth running and "drop" in RPM and that the RPM differential between magnetos does not exceed that specified in the Aircraft Flight Manual.

Ignition systems operate properly up to the point of maximum compression stroke pressure in the engine cylinders. This high pressure point is in the high RPM range, but may be well below maximum RPM. When an ignition system operates satisfactorily at maximum pressure, proper operation at lower pressures is ensured. Therefore, when other than the RPM specified in the Aircraft Flight Manual is used the check may not prove what it is supposed to prove.

Full Power Check

Most light aircraft engine run-up procedures no longer include a static full power check, but unless this is carried out nothing in the normal check will ensure that full power is actually available. Normally this check is carried out during the take-off roll, but if the surface allows, for a short field take-off do a full power check before allowing the aircraft to move.

Climbing

Most light aircraft climb at high power and relatively low airspeed compared with cruising flight. Since the engine is dependent upon the flow of outside air for cooling, the lower the airspeed the less effective the cooling. The normal climbing speed specified for an aircraft takes into account, among other things, the need for adequate cooling.

However, two other climbing speeds are generally specified for an individual aircraft:

1. Best angle of climb speed.
2. Best rate of climb speed.

As both of these climb speeds are often lower than normal climb speed, with resultant higher engine temperatures, their use should be limited to the period of time they are necessary, with normal climbing speed resumed as soon as possible.

In the case of engines designed to climb at full throttle until cruising altitude is reached, you gain no advantage by reducing power on climb with the thought of "sparing the engine" provided you strictly adhere to the specified climbing speed. One of the important reasons for maintaining the recommended speed concerns adequate cooling, which has already been discussed. Another reason is that with a fixed pitch propeller an airspeed higher than normal may cause engine RPM to exceed the limitations for sustained full throttle operation.

Unless otherwise specified, the procedure for aircraft equipped with a mixture control is to take off and climb with the control in the "full rich" position. Within certain bounds an aircraft engine runs cooler with a rich mixture, and since a power setting greater than that of normal cruise power generates much more undesirable heat the enriched mixture contributes greatly to the welfare of the engine.

Cruise Power

Most of the time the average light aircraft is operated in the normal cruising range. In determining the cruise power setting of an engine for a particular aircraft, the manufacturer strives for the best choice in consideration of reliability, performance, economy of operation, and engine life. Of these, engine reliability is the overriding factor. It goes without saying, therefore, that a sustained power setting in excess of that recommended for normal cruising may threaten a most important factor of safe flight, i.e., engine reliability.

General

When descending with low power settings or idle power, as in the case of a glide, an engine will cool rapidly even in relatively warm weather. A sudden application of power, such as for an overshoot, can damage a cold engine and/or result in a momentary engine malfunction. Therefore, it is good practice in a sustained descent to apply power periodically to retain engine operating temperatures.

Power should be increased and decreased by prompt but smooth operation of the throttle; this eliminates backfiring and the possibility of an abrupt loss of power at a crucial moment.

With normal use an engine will cool enough during approach, landing, and taxiing to permit shutting it off without further idling. However, if there has been an excessive amount of power used while taxiing, allow the engine to run two or three minutes at just above the idling speed, before you shut it off.

Prior to shutdown, at idle RPM, select the magnetos "off" momentarily to determine if the engine will stop. If it doesn't, you have a live magneto. This is very dangerous because the engine may fire should anyone turn the propeller by hand after shutdown. This condition should be reported immediately to the operator or an Aircraft Maintenance Engineer.

Aircraft Documentation

Your student pilot permit indicates that, subject to certain conditions, the holder may, for the purpose of their flight training and under supervision, act as pilot-in-command of any aircraft not carrying passengers. Pilot-in-command means the person responsible for successful completion of a safe flight. The pilot-in-command of any aircraft must be familiar with, and conform to, all the regulations and administrative requirements relating to the flight and in particular, the conditions under which the aircraft must be flown. In many respects these conditions are governed by the documents that must be carried on board an aircraft during flight.

Documents Required on Board

1. Student Pilot Permit, Pilot Permit or Pilot Licence and Medical Certificate.
2. Radiotelephone Operator's Restricted Certificate (Aeronautical).
3. Aircraft Radio Station Licence.
4. Certificate of Airworthiness or Flight Permit.
5. Certificate of Registration.
6. Aircraft Journey Log.
7. Copy of Liability Insurance (privately registered aircraft).
8. Aircraft Flight Manual (In most cases).

Document Information

It is the responsibility of the pilot-in-command to ensure that all documentation required for an aircraft and its crew is on board and valid for the flight.

Student Pilot Permit. Upon receipt of a satisfactory medical assessment, a Medical Certificate or Medical Assessment Letter will be issued by Transport Canada. This is required before the Student Pilot Permit can be validated. The Student Pilot Permit is normally issued and validated by an "Authorized Person" delegated by Transport Canada to carry out certain licensing actions. It shall be carried by the pilot while flying solo.

An expired Student Pilot Permit may be re-issued by satisfying all the requirements for initial issue as outlined in Part IV of the *Canadian Aviation Regulations.*

Pilot Permit or Pilot Licence. Pilots shall, during flight, carry a valid permit or licence, appropriate to their duties. This includes the Student Pilot Permit. The validity period of the permit or licence is determined from the accompanying Medical Certificate.

Regulations also require crew members to produce their permits or licences upon demand by designated Transport Canada officials, peace officers, or immigration officers.

The Medical Certificate, pilot permit and licence have important information on the back with which you must comply.

Radiotelephone Operator's Restricted Certificate (Aeronautical). Any person operating radio transmitting equipment installed in an aircraft registered in Canada is required to hold and carry this certificate issued by the Department of Communications.

Aircraft Radio Station Licence. All radio equipment used in Canadian registered civil aircraft require a valid radio station licence issued by the Department of Communications. The licence should be checked to ensure that the call sign is correct and the document shall be carried on board. The radio station licence must be renewed prior to the expiry date printed on the document.

Certificate of Airworthiness. The issuance of a Certificate of Airworthiness signifies that Transport

Canada is satisfied that the aircraft identified on the certificate conforms to Transport Canada recognized design standards and is considered fit and safe for flight on the date of issue of the certificate. A Certificate of Airworthiness makes the aircraft valid for flight in any ICAO Contracting State. For a Certificate of Airworthiness to remain in force, the aircraft must be safe and fit in all respects for the intended flight, and certain conditions must be met:

1. The weight of the aircraft and its load must not exceed the maximum permissible weight specified.
2. The load must be properly distributed.
3. The equipment and any cargo carried must be secured to prevent shifting in flight and placed to allow unrestricted exit of passengers in an emergency.
4. The required emergency equipment must be carried on board in good condition.
5. As well, to be safe and fit for the intended flight, aircraft owners and operators are responsible for compliance with all airworthiness directives (ADs) that are applicable to their aircraft. ADs are notices issued by Transport Canada concerning safety defects that must be rectified immediately or within a specified time period or flight time limitation.

   All aircraft maintenance requirements inspections must also be completed. These inspections may vary according to aircraft type, operation, or the maintenance program being followed.
6. Where an aircraft has undergone maintenance, the Certificate of Airworthiness or flight permit of that aircraft is not in force until a *maintenance release* has been entered in the aircraft journey log and other maintenance records and signed in respect of the work performed. The maintenance release indicates that the maintenance was performed in accordance with the applicable standards of airworthiness.
7. The Certificate of Airworthiness indicates under which category the aircraft may operate. Most operate under the standard category. When checking an older Certificate of Airworthiness, you may note the term normal, which means the same as standard. The Certificate of Airworthiness must be on board during flight, and the pilot must ensure that it is valid and the one issued for the aircraft.

A Special Certificate of Airworthiness is issued when an aircraft does not meet all the requirements for a standard Certificate of Airworthiness but is considered fit and safe for flight on the date of issue of the certificate. It is valid only in the country in which it was issued. A Special Certificate of Airworthiness may be issued in any one of the following classifications: provisional, restricted, amateur built, or lim-

ited. An agricultural spray aeroplane would be an example of an aircraft issued a Special Certificate of Airworthiness classified as restricted.

Flight Permit.   A Flight Permit may be issued for experimental or other specific purposes in place of a Certificate of Airworthiness.

An Experimental Flight Permit may be issued for any aircraft, excluding amateur built aircraft, manufactured for or engaged in aeronautical research and development, or for showing compliance with airworthiness standards.

A Specific Purpose Flight Permit may be issued for an aircraft that does not comply with applicable airworthiness standards but is capable of safe flight. It provides flight authority in circumstances when a Certificate of Airworthiness is invalidated or there is no other certificate or permit in force. A Specific Purpose Flight Permit may be issued for:

1. Ferry flights to a base for repairs or maintenance.
2. Importation or exportation flights.
3. Demonstration, market survey, or crew-training flights.
4. Test purposes following repair, modification, or maintenance.
5. Other purposes as determined by the Minister of Transport.

A Specific Purpose Flight Permit may carry restrictions with it such as a prohibition from carrying passengers; therefore, it should be checked carefully prior to operating the aircraft.

Certificate of Registration.   Every aircraft in Canada, other than hang gliders, is issued a Certificate of Registration. Except for balloons it shall be kept on board the aircraft during flight. Check the Certificate of Registration to ensure that it is the one issued for the aircraft. If a change of ownership is in process, there may be circumstances where the aircraft cannot be flown. Information concerning aircraft registration can be found in the Licensing, Registration, and Airworthiness section of the *Aeronautical Information Publication* (A.I.P.) Canada.

Aircraft Journey Log.   Prior to flight, ensure that the journey log is the one assigned to the aircraft and check that it is up-to-date. Check that all airworthiness inspections are completed and that the appropriate airworthiness entries and certifications have been made. Remember that irregularities in the journey log may invalidate the Certificate of Airworthiness. Close scrutiny of the journey log will assist the pilot in determining whether the aircraft is legal and safe to fly. Certain companies, including many flight training units, have been authorized to conduct specific types of flights without having to carry journey logs. Companies affected by this authorization have a letter from Transport Canada outlining the conditions under which the journey log need not be carried. It is recommended that copies of that letter be on board the aircraft in the event of inspection by a Transport Canada delegated official. If in doubt about whether the journey log must be on board, check with the operator before taking the aircraft.

Entries in the journey log shall be made in *ink* by a competent person, as soon as possible after the events occur. Should an error be made in an entry, do not erase or alter the entry. Draw a single line through the full length of the incorrect entry, initial it, and insert the correct entry in the next space. No person shall tear or remove any leaf from the log, or otherwise deface or destroy the log.

Every aircraft owner shall normally preserve each journey log maintained for a period of not less than one year after the date of the last entry in the log. The owner shall also, on the first page of every log taken into use to replace another log, make entries from the preceding volume necessary to ensure that an unbroken chronological record is maintained.

When making entries, it is important to differentiate between *flight time* and *air time*. Flight time is the total period of time from the moment the aircraft first moves under its own power for the purpose of taking off until the moment it comes to rest at the end of the flight. Air time is the time elapsed between the aircraft leaving the surface on take-off and touching the surface again on landing.

Liability Insurance. All Canadian aircraft, whether privately or commercially registered, are required to carry liability insurance. Privately registered aircraft must carry proof of insurance coverage on board during flight.

Aircraft Flight Manual. Information contained in this manual may be required by the pilot during the flight. Aircraft manufactured prior to the requirement for flight manuals may not be equipped with this document. In such cases the aircraft operating limitations must be conspicuously placarded in the aircraft. To be considered airworthy, some aircraft require that the Aircraft Flight Manual be carried as part of the documentation. Whether this requirement exists or not, if an Aircraft Flight Manual for the type and model to be flown is published, it is strongly recommended that it be carried at all times.

Other Documents

Weight and Balance. The specific weight and balance documents for the aircraft should be carried on board, especially when landing away from the pilot's home base.

Annual Airworthiness Information Report. A personalized Annual Airworthiness Information Report is sent to each registered aircraft owner normally five to six weeks in advance of the anniversary of the last issued Certificate of Airworthiness or Flight Permit. The aircraft owner shall complete the annual report by entering all data required and certifying that the information supplied is correct.

Submission of the annual report is mandatory, regardless of whether or not the Certificate of Airworthiness or Flight Permit is being renewed. It will *not*, however, cause the reinstatement of a previously expired or out-of-force Certificate of Airworthiness or Flight Permit.

One copy of the Annual Airworthiness Information Report is retained with the Certificate of Airworthiness or Flight Permit.

Interception Procedures. Every person operating an aircraft in Canadian airspace should understand the procedures for interception and visual signals. It is recommended a copy of these procedures be carried on board the aircraft. These can be found in Part VI of the *Canadian Aviation Regulations*, the Search and Rescue section of the *Aeronautical Information Publication* (A.I.P.) Canada, or the *Canada Flight Supplement.*

Aircraft Technical Log. Every aeroplane must have a Technical Log in which the overhaul and maintenance history of the aircraft's airframe, engine, propeller, components, and installations and modifications is recorded.

Pilot Log-book. You must maintain a pilot log-book of recognized form, with accurate, legible, certified entries. A personal log-book is a requirement for proof of experience for the issue of licences and endorsements throughout a pilot's career.

CHAPTER FIVE

Airport Operations

An aerodrome means any area of land, water (including the frozen surface thereof), or other supporting surface used, designed, prepared, equipped, or set apart for use either in whole or in part for the taking off, landing, surface manoeuvring, or servicing of aircraft.

In Canada there are two classes of aerodromes: aerodromes that are not certified as airports; and aerodromes that are certified as airports. Those certified as airports must be maintained and operated in accordance with applicable Transport Canada standards.

Aerodromes listed in the *Canada Flight Supplement* (CFS) that are not certified as airports are called registered aerodromes. Before using them, pilots should obtain current information on the aerodrome condition from the owner or operator. Where use of an aerodrome requires prior permission from the owner or operator the designator "PPR" is shown in the *Canada Flight Supplement*. However, in an emergency any airport or aerodrome may be used.

Notice to Airmen (NOTAM): Field Condition Reports

At many airports, the NOTAM office provides NOTAM and NOTAM summaries concerning operations that may affect an aircraft in flight at or in the vicinity of certain airports.

These NOTAM also contain field condition reports on various airports and should be reviewed very carefully by pilots prior to carrying out local or cross-country flights. Usually the airport NOTAM office is located in or adjacent to the Flight Service Station (FSS). Transport Canada's flight information publication, the *Canada Flight Supplement* provides information on those aerodromes where NOTAM service can be obtained and the telephone number of the operator where applicable.

When approaching an airport to land, never hesitate to ask the control tower (or Flight Service Station) for a field condition report if you have any concerns, especially when winter airport maintenance is being carried out.

Runway Conditions

Snow, slush, or standing water on a paved runway can seriously degrade take-off performance. The length of grass or crops, gravel, roughness of the surface, or mud on any unprepared area selected for take-off can also greatly affect aircraft performance. Aircraft Flight Manuals often give an approximation of the increase in take-off length under some of these conditions. Due to the variables involved, such as depth of snow, no one figure can be applied to all situations.

Pilots must take great care in assessing their own capabilities and the capabilities of their aircraft when deciding whether to attempt a take-off under any one or combination of these conditions. It is recommended that pilots select a reject point somewhere along the take-off path. If the aircraft has not lifted off by this point, the pilot should throttle back and stop. The reject point should take into consideration adequate room to stop if required to do so and also the clearance of obstacles off the departure end once airborne. Pilots will often walk the runway to check it before making a final decision. If in doubt, don't go!

When landing at an unfamiliar location, it is important to obtain prior information concerning the condition of the runway or landing area. A couple of inches of snow could be enough to put an aircraft on its nose or its back. Ice or standing water could cause the aircraft to have reduced or no braking action which could be disastrous if the length of the landing area is short. Prior information may not be available or perhaps conditions changed en route. Upon arrival it is extremely important that the pilot check the landing area carefully to determine its suitability. The pilot must be prepared to divert elsewhere if it is doubtful that a landing can be accomplished. This contingency should be included in the pre-flight planning.

Surface Winds

Control towers and Flight Service Stations broadcast surface winds for the airport where they are located, in knots and in degrees magnetic. As well, a Flight Service Station providing a Remote Aerodrome Advisory Service (RAAS) through a Remote Communications Outlet (RCO) will broadcast surface winds in knots and degrees magnetic. At sites other than the airport where the Flight Service Station is located, or providing Remote Aerodrome Advisory Service, winds are given in degrees true.

Wind Direction Indicator

At aerodromes that do not have prepared runways, the wind direction indicator is usually mounted on or near some conspicuous building associated with the aerodrome or the general aircraft parking area.

Wind direction indicators may be located near both ends of long runways. At aerodromes with shorter runways the wind direction indicator will be centrally located so as to be visible from approaches and the aircraft parking area. For night operations the wind direction indicator will be lighted.

At aerodromes certified by Transport Canada for public use, a dry standard wind direction indicator will react to wind speed as follows:

Wind Speed	Wind Indicator Angle
15 KT+	Horizontal
10 KT	5 degrees below horizontal
6 KT	30 degrees below horizontal

At other aerodromes, non-standard wind indicator systems may be in use that could react differently to wind speed.

Manoeuvring Surfaces

An aircraft should not be operated on the aircraft manoeuvring surfaces at a controlled airport without a clearance to do so from the control tower. Even if the aircraft is being moved for purposes other than intended flight, the control tower must be advised by radio, telephone, or other means.

Taxiway centre-line markings and holding-point markings are readily distinguished from runway markings. The colour used is yellow instead of white and they are a relatively narrow width of 6 inches. Taxiway holding-point markings consist of solid and broken lines across the taxiway (or holding bay) parallel to the runway (Fig. 1-24).

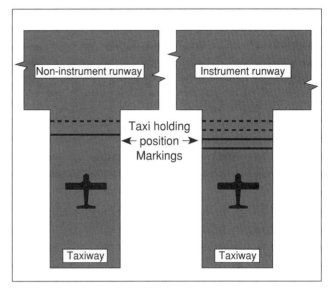

Figure 1-24 Taxi Holding Position

Taxi Holding Positions

Taxi holding positions are established at many airports, but at airports where they have not been established an aircraft should be held at least 200 feet (61 metres) from the edge of the runway in use. Where this is not possible, aircraft should be held at a distance that does not create a hazard to arriving or departing aircraft.

Runway Button

The term "button" has no official definition. It normally means the point at the end of a runway where an aircraft is positioned for take-off with the intention of having the full length of the runway available.

Runway Numbering

Runways are numbered according to their direction. A runway running east and west would be called runway 09-27. These numbers represent the bearings of 090 and 270 degrees with the last digit removed. The number nearest the even 10 degrees division is used. For example, a runway bearing 134 degrees is numbered 13 at one end and 31 at the other. If an aircraft is aligned with runway 14, the compass should indicate a heading between 135 and 144 degrees.

Runway Threshold Markings

The markings for runway thresholds consist of lines across the width of the runway at the threshold ends parallel with the runway centre line.

The threshold of a runway may be temporarily or permanently displaced to take a poor surface out of service or to bring the runway within zoning standards. Under certain conditions you may be tempted to land before the displacement marking because the displaced part may look reasonably serviceable. Avoid this. The area may not receive regular maintenance and could contain hazards not readily visible, especially during the winter when windrows of snow may be left by snow-ploughs.

The markings that indicate a temporarily displaced threshold consist of a white line placed across the width of the runway to indicate its new end, and four white arrowheads equally spaced across the width of the runway on the approach side of the runway end with their points indicating the new threshold. A permanently displaced threshold will have a white line across the width of the runway and two white arrows on the approach side, one following the other on the former centre-line, leading up to and pointing toward the new runway threshold.

Centre-Line Markings

Centre-line markings of a runway are useful both as a directional guide and as an aid to depth perception because of the "broken" line presentation. The marks are 100 feet long and longitudinally spaced 100 feet apart at airports under Transport Canada's jurisdiction.

Responsibility

Whether an airport is controlled or uncontrolled, nothing relieves the pilot-in-command of the responsibility for exercising good judgement and adhering to the rules of the air when an aircraft is being manoeuvred on or in the vicinity of an airport. At a controlled airport, air traffic control will provide assistance to increase the safety of the operation, but the pilot-in-command must never assume that operation of the aircraft is handed over to that agency. For example, when a tower controller clears an aircraft across a runway or through any active area, the pilot-in-command must still check that there are no hazards before proceeding.

CHAPTER SIX

Weather Considerations

Encountering Weather Below VFR Conditions

If you are qualified for flight under the Visual Flight Rules (VFR) only, plan your flight so that there is no risk of encountering weather below these conditions. However, always have an alternate plan should the weather deteriorate. Normally, the best plan is an early 180 degree turn to fly back into better weather. When no alternate appears available and the weather continues to deteriorate, contact ATC. They will do all that is possible to:

1. Provide information concerning an alternate route and radar navigation assistance if this will enable the flight to be continued in VFR weather conditions; or
2. If the above action is not practicable, provide radar navigation or radar approach guidance. If you have any doubt about the safety of your flight, don't allow a doubtful situation to develop into a bad one. Declare an emergency.

Special VFR

When weather conditions are below VFR minima, Special VFR flight in a Control Zone may be authorized by the appropriate ATC unit subject to current or anticipated Instrument Flight Rules (IFR) traffic.

An ATC unit cannot suggest Special VFR to a pilot. If the controller or Flight Service Station operator indicates that the weather is below VFR, it becomes the responsibility of the pilot to request Special VFR. Authorization is normally obtained through the local tower or Flight Service Station and must be obtained prior to operating in or entering a Control Zone. It does not relieve pilots from the responsibility of avoiding other aircraft, obstructions, or weather conditions beyond their capabilities.

Pilots should assess their abilities and the capabilities of their aircraft very carefully before intentionally flying in Special VFR conditions. Remember weather is never a constant and can change quickly.

White-out

Each winter there are a number of aircraft accidents as a result of pilots flying into white-out conditions and becoming disoriented due to reduced visibility, the lack of distinguishable features on the ground, and the loss of a visual horizon. The causes and effect of the phenomenon known as white-out may be described as follows:

Overcast White-out. A product of a uniform layer of cloud over a snow-covered surface. The rays from the sun are scattered and diffused as they pass through the cloud and are then reflected by the snow surface in all directions. As a result, the space between the ground and cloud appears to be filled with a diffused light with a uniform white glow. Depth perception is completely lacking as the sky blends imperceptibly with the ground at the horizon line, causing disorientation.

Water Fog White-out. Produced by the clouds containing supercooled water droplets with the cloud base usually in contact with the cold snow surface. Visibility both horizontally and vertically is affected by the size and distribution of the water droplets suspended in the air.

Blowing Snow White-out. Produced by fine blowing snow plucked from the snow surface and suspended in the air by winds of 20 KT or more. The suspended grains of snow reflect and diffuse sunlight and reduce visibility.

Precipitation White-out.   Although all falling snow reduces visibility, small wind-driven snow crystals falling from low clouds above which the sun is shining produce a white-out condition. The multiple reflection of light between the snow-covered surface and the cloud base is further complicated by the spectral reflection from the snowflakes and the obscuring of the landmarks by the falling snow.

PART TWO

AIR EXERCISES

Familiarization

The first flight will involve very little formal instruction. You will occupy the seat from which you will subsequently fly the aircraft. Your role as student will be mainly that of an observer. This flight will begin to accustom you to the sensation of flying and to the appearance of the countryside from the air. Your flight instructor may also include Exercise 4, "Taxiing," and Exercise 5, "Attitudes and Movements."

The first flight may be an entirely new experience. Remember, what looks complicated and difficult at this time will become less so as your flight training progresses.

You may be asked to keep your hands lightly on the controls and your feet lightly on the rudder pedals. The instructor will emphasize that only small, smooth control movements are required to control the aircraft and will briefly discuss the procedures to be followed in future flight training exercises.

Do not hesitate to ask questions. The instructor's voice must be completely audible and understandable; if it is not, say so.

The flight instructor will point out readily identifiable local landmarks and explain their orientation to the airport. The function of various flight instruments will also be explained, and from time to time you may be asked to state the altitude and the speed of the aeroplane.

You will also be given the responsibility to assist in looking out for other aircraft that may be in the vicinity. These aircraft could be in any position relative to you and flying in any direction. Their position should be passed on immediately to the instructor, who will decide if any avoiding action is necessary.

Aircraft Familiarization and Preparation for Flight

This exercise begins with the flight instructor acquainting you with the type of aircraft to be used during the training period. The main components of the aircraft will be pointed out and the function of each carefully explained. For example:

Wings.   The wings provide the lift required to make the aeroplane fly by obtaining a useful reaction from the air through which the wing moves.

Fuselage.   This is the main body of the aircraft. It is what is left if the wings, engine, landing gear, and tail surfaces are removed.

Tail Surfaces.   These may be separated into horizontal stabilizer (tail plane) and vertical stabilizer (fin).

As the names imply, they provide the aircraft with stability in certain planes of movement.

Ailerons, Elevators, and Rudder.   These are movable aerofoil surfaces, which enable the pilot to manoeuvre and control the aircraft in flight.

The ailerons are positioned on hinges toward the outer ends of the wings; one moves up as the other moves down when the control column is moved from side to side. They are used to control bank in flight.

The elevators are hinged to the trailing edge of the horizontal stabilizer and are moved up or down when the control column is moved backward or forward. They are used to pitch the aeroplane up or down in flight.

The rudder is hinged to the trailing edge of the vertical stabilizer and is linked to the rudder pedals. The pilot controls yaw by means of the rudder.

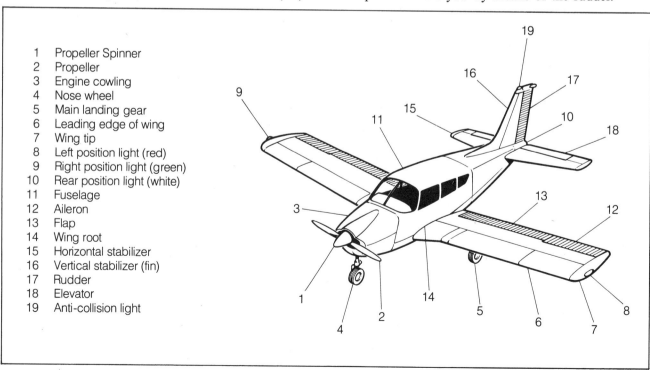

1	Propeller Spinner
2	Propeller
3	Engine cowling
4	Nose wheel
5	Main landing gear
6	Leading edge of wing
7	Wing tip
8	Left position light (red)
9	Right position light (green)
10	Rear position light (white)
11	Fuselage
12	Aileron
13	Flap
14	Wing root
15	Horizontal stabilizer
16	Vertical stabilizer (fin)
17	Rudder
18	Elevator
19	Anti-collision light

Figure 2-1   Major Components of an Aircraft

Landing Gear.   The function of this component requires little explanation, other than that it includes at least the supporting struts, wheels, tires, and any springs or other devices to absorb shocks due to uneven terrain or landing impact. Floats and skis are components of other types of landing gear.

Flaps.   These are movable surfaces forming part of the wing and are mounted at or near the trailing edge, between the wing root and the ailerons. They extend and retract together and increase or decrease the effective lift of the wing by altering its camber and in some cases the area. They are controlled by the pilot.

Trim Tabs.   These are small ancillary aerofoils hinged to the elevators and, in some aircraft, to the ailerons and rudder. They help the pilot by reducing control pressures induced by changing flight attitudes.

Flight Controls.   These consist of the control column and the rudder pedals. The control column may consist of a wheel arrangement or a straight "stick." Rotate the wheel (or move the stick) to the left, and the left aileron moves up while the right aileron moves down; rotate the wheel (or move the stick) to the right, and the right aileron moves up while the left aileron moves down. Push the left rudder pedal, and the rudder moves to the left; push the right rudder pedal, and the rudder moves to the right. The rudder pedals may also be used to "steer" the aircraft on the ground by means of a steerable nose wheel or tail wheel; pushing the left rudder pedal turns the aircraft to the left, and vice versa.

Throttle.   This is the power control. To increase power, move the throttle forward; to decrease power, move the throttle back. Most throttle arrangements include a device for increasing or decreasing tension on the throttle so that the power setting does not change when your hand is removed from the throttle.

Ancillary Controls.   These comprise the mixture control and carburettor heat control (see Exercise 3, "Ancillary Controls").

Environmental Controls.   These are normally the windshield defogger, and the heating and ventilation controls (see Exercise 3, "Environmental Controls").

Instrument Panel.   The instrument layout directly in front of the pilot is called the instrument panel. The instruments may be divided into three categories: *flight instruments*, *engine instruments*, and *navigation instruments*.

Flight Instruments

Airspeed Indicator.   This instrument indicates the aircraft's speed through the air. It relates only indirectly to the speed of the aircraft over the ground. It may indicate speed in miles per hour (mph) or knots (KT). Knots are nautical miles per hour.

Altimeter.   This is a pressure sensitive instrument which, if properly set, indicates the height at which the aircraft is flying. The customary procedure is to set the instrument so that it indicates height above mean sea level. When the aircraft is on the ground, the indication on the altimeter will be that of the elevation of the aerodrome.

Turn-and-Bank Indicator.   The needle portion of this instrument indicates whether the aircraft is turning, together with the direction and rate of turn. The ball portion of the instrument is fundamentally a reference for co-ordination of controls. In all co-ordinated flight the ball will be centred in its curved glass tube. Instead of a turn-and-bank indicator, the aircraft may be equipped with a turn co-ordinator, which provides basically the same information, but with a different display.

Magnetic Compass.   This is the basic reference for heading information. The compass points toward the magnetic north pole, but it is susceptible to certain errors such as oscillations in turbulence and incorrect readings during turns or when influenced by other magnetic attractions. These errors are explained in Exercise 6, "Straight-and-Level Flight."

Heading Indicator.   This gyroscopic instrument has no magnetic qualities of its own and, therefore, must be set periodically by reference to the magnetic compass. Its main asset is that it provides a stable directional reference, and (unlike the magnetic compass) is relatively free of error during turns, acceleration, and deceleration in normal flight manoeuvres.

Attitude Indicator.   This gyroscopic instrument is an artificial horizon. The miniature aircraft superimposed on its face enables the pilot to determine the aircraft's attitude relative to the real horizon.

Vertical Speed Indicator.   This pressure sensitive instrument indicates the rate at which the aircraft is climbing or descending, in feet per minute.

Outside Air Temperature Gauge.   This is not a flight instrument, but it is a valuable aid to flight safety since its indications can help the pilot assess the possibility of icing conditions, including carburettor ice. The instrument usually registers outside air temperature in both degrees Fahrenheit and degrees Celsius.

Engine Instruments

Tachometer.   This instrument indicates the speed at which the engine crankshaft is rotating in revolutions per minute (RPM). In aircraft with a fixed pitch propeller, RPM is directly related to power. As the throttle is eased forward, the tachometer will indicate an

increase in RPM. As the throttle is eased back, the opposite effect is achieved.

Oil Pressure Gauge.   This vitally important instrument registers, usually in pounds per square inch, the pressure of the lubricating oil being supplied to the engine. Refer to the Aircraft Flight Manual for limits and recommended pressures during particular phases of engine operation.

Oil Temperature Gauge.   This gauge indicates the temperature of the engine lubricating oil. It reacts by showing higher than normal temperatures if the oil pressure system malfunctions or if engine cooling is inadequate.

Airspeed indicator   Attitude indicator   Altimeter

Turn-and-bank indicator   Heading indicator   Vertical speed indicator

Magnetic compass   Turn co-ordinator

Figure 2-2   Basic Flight Instruments

Manifold Pressure Gauge. This instrument indicates the pressure in the intake manifold of the engine, which relates to the pressure developed for any given condition of throttle setting and RPM. It is mainly used on aircraft equipped with a constant speed propeller. This instrument is calibrated in inches of mercury.

Navigation Instruments

VOR (Very High Frequency (VHF) Omnidirectional Range) Receiver Display and Control Unit. This instrument permits an aircraft to track to or from a VOR ground station on any track the pilot selects.

ADF (Automatic Direction-Finder). The ADF display includes a pointer (often referred to as the "needle") that points in the direction of any suitable ground radio station tuned into the ADF receiver by the pilot.

GPS (Global Positioning System) Receiver. By utilizing information received from GPS satellites, the GPS receiver displays accurate position information allowing an aircraft to fly a direct track to any point on the Earth. The receiver can also calculate a variety of navigation solutions, such as, groundspeed, track, and time to destination.

Fuel System

The fuel system of the average training aircraft is relatively simple, but the fact that the fuel supply to the engine is under the pilot's control requires that it be understood thoroughly. The fuel system must be managed as outlined in the Aircraft Flight Manual.

Electrical System

The electrical system of a light aircraft is not a great deal different in principle from that of an automobile: there is a generator or alternator, a starter, a battery, a voltage regulator, an ammeter, and various switches for carrying out routine functions.

Master Switch. When the master switch is turned on, it connects the battery to the electrical system. Electrical power is required to operate specific instruments and equipment and is also used to start the engine. Once the engine is started, it runs independently from the electrical system, requiring only the magnetos for its operation. If the master switch is turned off, you will lose all electrically powered instruments and equipment, but the engine will continue to run.

Ignition System

The engine ignition system uses magnetos for its electrical power source. As an added safety feature, each engine has two magnetos, and each cylinder has two spark plugs firing simultaneously. Should a magneto fail in flight, the engine will run quite satisfactorily on the one remaining, with only a slight loss of power.

Communications System

The air-to-ground communications system usually consists of a two-way radio operating on VHF. Although two-way, it is not simultaneous like a telephone. The pilot presses a button on the microphone to talk, and during this period all other nearby transmitters and receivers on the same frequency are jammed. The pilot must release the microphone button after the transmission to receive a reply. Transmissions must be short and to the point.

Flight Preparation

It is good practice to take all the charts and publications required for the proper navigation of the flight. No VFR flight should leave the ground, regardless of how short its duration, without current aeronautical charts on board covering the area in which the flight will be conducted.

After releasing the aircraft tie-downs, ensure that the following actions are carried out:

1. Remove exterior flight control locks, where applicable.
2. Remove pitot tube cover.
3. Remove wheel chocks.
4. Free flight controls of any locking or securing system in the cockpit.

The pre-flight external line check, often referred to as the "walk around," determines from the pilot's point of view that the aircraft is serviceable and that it has sufficient fuel and oil for the intended flight. A recommended line check is included in most Aircraft Flight Manuals.

Your first impression of an aircraft cockpit may be that it is cramped and complex. This impression will quickly disappear as you learn the proper method of entering the cockpit and adjusting the seat and controls. Once comfortably seated, fasten the seat-belt and shoulder harness and learn how to adjust and release them. Fastening the seat-belt and shoulder harness as soon as possible after you are seated in the cockpit is a habit to acquire immediately.

In a seaplane operation, the practice concerning seat-belts and shoulder harnesses is slightly different. It is considered good seaplane practice to delay fastening seat-belts and shoulder harnesses until ready for take-off and to release them when the aircraft reaches the taxiing mode following a landing.

The flight instructor will explain the opening and proper securing of doors, panels, and windows. Cabin and baggage door latches used on most small aircraft do not provide a conspicuous visual indication that a door is not secure, nor are such aircraft usually equipped with a "door open" warning device in the cockpit. Therefore, if a pilot becomes distracted or is otherwise inattentive when performing the pre-flight checks, an unsecured door can be overlooked. The unexpected opening of a door, window, or panel during any phase of flight can be very disconcerting due to noise and airframe buffeting. In some aircraft, control may be affected by increased drag, loss of lift, and adverse aircraft stability. (See Exercise 29.)

The aircraft should be positioned so that the propeller slipstream does not present a hazard or nuisance to others when the engine is started or during the subsequent engine run-up. It is discourteous and thoughtless to start an engine with the tail of the aircraft pointed toward a hangar door, parked aircraft or automobiles, or a crowd of spectators. The ground or surface under the propeller should be firm, smooth turf or concrete if possible, so that the propeller slipstream does not pick up pebbles, dirt, mud, or other loose particles and hurl them backward, damaging not only the rear portions of the aircraft but often the propeller itself. Avoid blocking a taxiway.

Before starting the engine, make a geographic check of all items in the cockpit, and then do a pre-start check. These checks are very important. For example, if the brakes are not on and secure when the engine starts, the aircraft may move forward unexpectedly. Starting the engine with the carburettor heat on can damage the carburettor heat system should the engine backfire. The battery may become discharged should you vainly attempt to start the engine with the mixture control in the idle cut-off position or with the fuel selector valve in the "off" position.

Your flight instructor will show you the procedures that apply to your own aircraft. The pre-take-off checks are carried out as specified in the Aircraft Flight Manual. One item important to remember is that if there is to be a change in fuel tanks before take-off, make it before the engine run-up. Then, if a fuel malfunction exists, it will show up before take-off.

Parking

Carefully select the surface upon which to park or stop an aircraft. Avoid icy surfaces, since the wind exerts considerable force upon an aircraft and may move it even though the brakes are securely applied. During engine start or run-up, the wheels must not be on a slippery surface, otherwise the aircraft may move forward against its brakes. When parking the aircraft make sure it is left on firm ground, otherwise it may settle and subsequently prove very difficult to move under its own power.

If for any reason the brakes have been used a lot, the braking system may have been overheated. Should the parking brake be applied immediately under these circumstances, the brakes may self-release when the system cools. When overheated brakes are suspected, allow a suitable time interval for cooling to take place before you apply the parking brake, or chock the two main wheels, both ahead and behind each wheel, before leaving the aircraft.

When an aeroplane is parked for any length of time, or if high winds exist or are forecast, it should be *tied down*. Tie-downs consist of appropriate lengths of rope or nylon line attached to weights, pickets, or other devices on or in the ground, with which the aircraft may be securely tied. The correct way to tie down a specific aeroplane is usually outlined in the Aircraft's Flight Manual.

In addition to the tie-down, the aircraft being parked should also be secured as follows:

1. Secure flight controls in cockpit.
2. Set the parking brake. For additional security, place wheel chocks ahead and behind each main wheel.
3. Install the pitot tube cover.
4. Install the exterior control locks, where applicable.

Surface Contamination

Never attempt to take off with any frost, ice, snow, or other such contaminants on any of the critical surfaces. Critical surfaces are the wings, control surfaces, propellers, upper surface of the fuselage on aircraft that have rear mounted engines, horizontal stabilizers, vertical stabilizers, or any other stabilizing surface of an aircraft. Even a minute film of ice or frost can seriously reduce the lift qualities of an aerofoil. It has been demonstrated that contamination having the thickness and roughness of medium to coarse sandpaper can reduce lift by as much as 30 percent and increase drag by 40 percent. Insofar as frost, ice, or snow on other parts of the aircraft are concerned, consideration must be given to the additional load and the possibility of the controls becoming jammed due to chunks dislodging or the refreezing of melting precipitation.

Removal of the contaminant by cleaning or de-icing can be accomplished in many ways; using squeegees, brushes, brooms, etc., or by using approved de-icing fluids containing various mixtures of hot water and glycol depending on the conditions.

De-icing fluids with higher concentrations of glycol may provide some protection against refreezing, but do not provide much protection against further accumulation. Accumulations of mud can be removed through washing. In winter conditions, care must be taken to ensure that the water used does not result in a covering of ice.

Only close inspection can determine if an aircraft is clean, and the inspection should be performed just prior to take-off if there is a possibility that the aircraft may be contaminated. In some cases while taxiing, water, snow, and mud could be thrown up onto the wings and tail surfaces by the wheels, propeller(s), or other aircraft. If the pilot cannot confirm that the aircraft is clean, the take-off must not be attempted. It is the ultimate responsibility of the pilot-in-command to ensure that the aircraft is in a condition for safe flight.

Survival Equipment and Clothing

Because the average light aircraft is as comfortable and warm as an automobile, there may be a temptation to treat the need for proper clothing and survival equipment too lightly. Should you be forced to make an unscheduled landing, there is every possibility that the landing site will be in a remote or isolated area where warmth and shelter are not immediately at hand. Proper clothing and equipment at such a time are essential to your welfare, perhaps even to your survival. Due to the varying climate and terrain of a country as vast as Canada it is difficult to specify all requirements; however, you can use as a guide the equipment and clothing listed under the following subject headings in Transport Canada's *Aeronautical Information Publication*: "Sparsely Settled Areas," "ELT (Emergency Locator Transmitter)," "Life-Saving Equipment," and "Single-Engine Aircraft Operating in Northern Canada."

Emergency Locator Transmitter (ELT)

An ELT is a unit carried on board an aircraft which, when activated, transmits a distinctive audible signal on the emergency frequencies of 121.5 megahertz (MHz) and 243.0 kilohertz (KHz). This enables search and rescue personnel to quickly locate downed aircraft.

As there are many types on the market, you should be familiar with how to operate the ELT in the aircraft you are flying. Details on the need to carry an ELT and its testing and use may be found in the Search and Rescue section of the *Aeronautical Information Publication*.

Passenger Safety Briefing

If you are carrying passengers, a safety briefing is essential. They should know about the use of seatbelts and shoulder harnesses, smoking limitations, and how the doors work. They should know what actions to expect in the event of an emergency landing. To be prepared for this, they must know the location and use of emergency exits, the ELT, the fire extinguisher, the first-aid kit, and other items for use in an emergency. Items specific to the aircraft being used (e.g. life vests) must be included in the briefing.

Ancillary Controls

Although the usual definition of *ancillary* is "subordinate to" or "auxiliary," ancillary controls are vital to the safe and comfortable operation of an aircraft. The ancillary controls to be discussed here are the carburettor heat control, the mixture control, and the environmental controls.

Carburettor Heat Control

Icing

Under certain moist (and "moist" is a key word) atmospheric conditions, with air temperatures ranging anywhere from −13°C to +38°C, it is possible for ice to form in the induction system (Fig. 2-3). The rapid cooling in the induction system using a float type carburettor is caused by the absorption of heat from the air during vaporization of the fuel and is also due in part to the high expansion of air through the carburettor venturi. As a result of the latter two influences, the temperature in the venturi may drop as much as 21°C below the temperature of the incoming air. If this air contains a large amount of moisture, the cooling process can cause precipitation in the form of ice, which may build up to such an extent that a drop in power output results and, if not corrected, may cause complete engine stoppage. Indications of icing to the pilot are a loss of RPM with a fixed pitch propeller, and a loss of manifold pressure with a constant speed propeller, together with the accompanying airspeed loss and engine roughness with both types.

To counteract the formation of carburettor ice, an aircraft is equipped with a controllable system for preheating the air before it enters the carburettor.

Carburettor Heat

Always anticipate possible icing and use carburettor heat before the ice forms. However, should ice begin

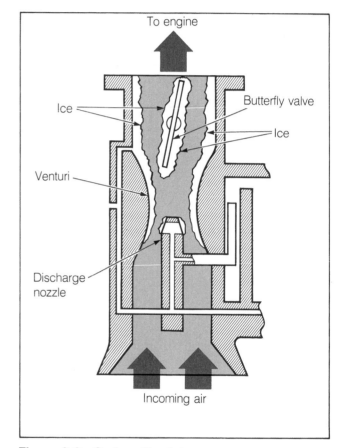

Figure 2-3   Carburettor Icing

to form, use the "full hot" position long enough to be sure of eliminating the ice. Using full heat will initially cause a loss of power and possible engine roughness. Heated air directed into the induction system will melt the ice, which goes through the engine as water, causing some of the roughness and more power loss. Despite this temporary roughness and power loss, a pilot is not damaging the engine at a cruise power of 75 percent or less with any amount of heat.

When using carburettor heat, there are related factors to remember. The engine loses an average of 9 percent of its power when full heat is applied, due to the reduced volumetric efficiency of heated air and loss of the ram air feature. Carburettor heat also creates a richer mixture, which may cause the engine to run rough, particularly in the full hot position. If there is any throttle available, bring the power up to the former RPM setting, then lean out and readjust the mixture until the engine runs smoothly again. Readjust the mixture as subsequent throttle and carburettor heat changes are made.

At low power such as in the traffic pattern, it may not be practical to lean the mixture.

Carburettor icing may be controlled or avoided by adopting the following practices:

1. Start the engine with the carburettor heat control in the "cold" position, to avoid damage to the carburettor heat system.

2. When relative humidity is high and the summer ambient temperature is below 28°C, use carburettor heat immediately before take-off. In general, carburettor heat should not be used while taxiing because in the "on" position, intake air usually bypasses the carburettor air filter.

3. Avoid using carburettor heat during take-off since it may cause detonation and possible engine damage. An exception to this might be in very low temperature areas, which call for special procedures.

CARB ICING

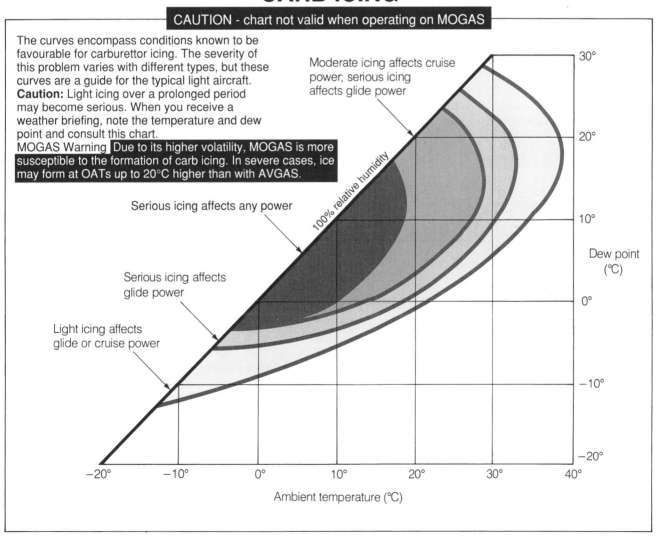

Figure 2-4   Carburettor Icing Graph

4. Remain alert after take-off for indications of carburettor icing, especially when visible moisture is present.

5. With a carburettor air temperature gauge, partial carburettor heat should be used as necessary to maintain safe temperatures to forestall icing. Without such instrumentation, use full heat, if you consider carburettor heat necessary.

6. When carburettor ice is suspected, immediately apply full heat. Watch for a power loss to indicate the presence of carburettor heat, then an increase in power as ice melts.

7. If carburettor ice persists after a period of full heat, gradually increase power to obtain the greatest amount of carburettor heat.

8. Carburettor icing can occur with the ambient temperature as high as +38°C and humidity as low as 50 percent. Remain especially alert with a combination of ambient temperature below +28°C and high relative humidity. The possibility of carburettor ice decreases (a) in the range below 0°C, because of lessened humidity as the temperature decreases, and (b) at around −10°C because of ice crystals that pass through the induction system harmlessly. It should be remembered that if the intake air does contain ice crystals, carburettor heat might actually cause carburettor icing by melting the crystals and raising the moisture laden air to the icing temperature.

9. During descents when carburettor icing is present or suspected, apply full carburettor heat and periodically apply sufficient power so that enough engine heat is produced to prevent or disperse ice. This is a general rule for many aircraft. Consult the Aircraft Flight Manual for the procedures in a specific aircraft.

The diagram of a carburettor heat system (Fig. 2-5) shows that when the system is in the "heat on" mode, air entering the carburettor is no longer filtered. This is the main reason for ensuring that the system is in the "heat off" mode when taxiing. At ground level the air may be laden with airborne particles harmful to the engine if ingested; at 100 feet above ground level this possibility is almost negligible.

There is a misconception that it does not matter to the efficiency of the engine whether the carburettor heat is on or off. If this were true, engine manufacturers would design their engines so that heated air was constantly directed through the carburettor air intake system, to completely eradicate the problem of carburettor icing. But they don't, because the application of carburettor heat in standard atmospheric conditions will:

1. Reduce the maximum power output of the engine.
2. Increase fuel consumption.

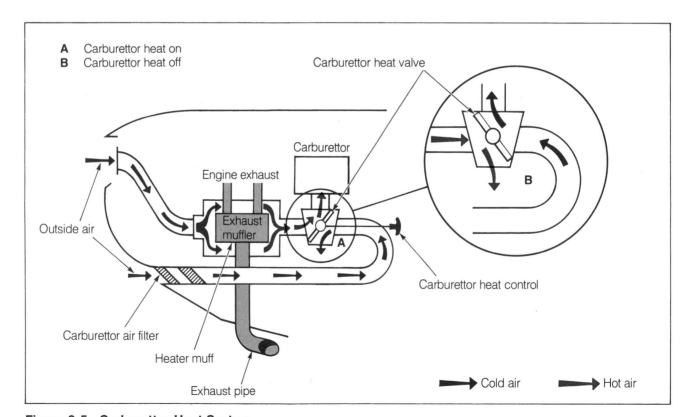

Figure 2-5   Carburettor Heat System

As the ambient temperature decreases, the effect of carburettor heat on the efficiency of the engine also decreases. Light aircraft engines operated at extremely low (winter) ambient temperatures may require the warming influence of carburettor heat to ensure adequate response to throttle application.

Mixture Control

As an aircraft gains altitude, the surrounding air becomes less dense. Atmospheric pressure of approximately 14.7 pounds per square inch at sea level is only 10.2 pounds per square inch at 10,000 feet above sea level. At altitude, the engine draws a lesser weight of air into its cylinders than it does on the ground. If the weight of the fuel drawn into the cylinders remained the same, regardless of altitude, the mixture would become too rich at altitude.

Since the carburettor of an aircraft engine is adjusted to give maximum power for take-off, as the aircraft gains height the fuel/air mixture gradually becomes too rich. This imbalance may be corrected by use of the mixture control, which is a lever or push-pull knob within easy reach of the pilot.

To obtain the best performance from an engine a proper fuel/air mixture is important. Unless otherwise specified, the customary procedure is to take off and climb with the control in the "full rich" position, and leave it there while operating within the airport traffic pattern. Within certain bounds an aircraft engine runs cooler with a rich mixture, and since a power setting greater than that of normal cruising power generates much more undesirable heat, the enriched mixture contributes to the welfare of the engine. Some Aircraft Flight Manuals suggest leaving the mixture control lever in the "full rich" position below specific altitudes. Only on rare occasions would mixture control have to be leaned for take-off and climb, for example at airports with a very high elevation. Some aerodromes on the North American continent have field elevations of over 5,000 feet.

Correct procedures for mixture control are outlined in the Aircraft Flight Manual. However, the generally accepted procedure for leaning the mixture is to move the mixture control slowly toward the "lean" position until maximum RPM is obtained with a fixed power setting. Then, move the mixture control toward "rich" until a decrease in RPM is just perceptible. This produces optimum power for the throttle setting, with a slightly rich mixture to prevent overheating, since sustained operation with the mixture too lean can damage the engine. It is important to note that aircraft performance chart fuel flow and range figures are based

on the recommended lean mixture setting detailed in the Aircraft Flight Manual.

Many light aircraft pilots place the mixture control in the "full rich" position prior to a routine descent from altitude. The prime reasons for doing this are to ensure that the mixture will not be too lean should there be a sudden need for power and to guarantee that the control is in the proper position for the approach to landing. Ideally, the mixture should be adjusted gradually toward "full rich" as the descent progresses. An overrich mixture tends to cool an engine, which is probably already being overcooled due to the lower power-to-speed ratio of the descent itself. Prolonged descents with the mixture in the "rich" position have been cited by some engine manufacturers as one of the causes of premature spark plug failure.

The correct adjustment of the mixture control and a technical knowledge of how it affects the engine, plus its overall effect on the total aircraft operation are most important, especially during cross-country flights in a varying environment. For example, when the fuel/air mixture is too rich:

1. The engine may not develop the rated power.
2. The engine will run unevenly.
3. The engine may operate cooler than is desirable.
4. Fuel is wasted.
5. There is increased possibility of spark plug fouling.
6. Range is reduced.

On the other hand, when the fuel/air ratio is too lean:

1. Power will be lost.
2. The engine may run roughly and be subject to unnecessary vibration.
3. The engine may operate hotter than is desirable.
4. The engine may be damaged due to detonation.

Under normal flight conditions, if an aircraft is not equipped with fuel/air ratio instruments, it is best to adjust the mixture so that it is on the rich side.

Environmental Controls

Windshield Defogger

At all times the windshield of an aircraft should be kept clear and free of anything that will interfere with forward visibility, not only for control purposes but also to see outside fixed obstructions and other air traffic clearly. For those aircraft so equipped, the windshield defogging system will generally keep the windshield clear of interior fogging when the aircraft is in flight; however, while taxiing or waiting at the take-off position, or during run-ups, windshield fogging may occur. On these occasions, fogging may be controlled by opening the aircraft door or window slightly to improve interior air circulation. This procedure has also been found effective for aircraft without windshield defoggers.

Take care to ensure that the door and window are closed and properly latched again before take-off.

Heating and Ventilation

The cabin heating arrangement of most training aircraft consists of a controllable system of directing outside *ram* air (air entering an air inlet as a result of the forward motion of the aircraft), through an exhaust manifold heat exchanger and by flexible tubing into the cabin area. Many aircraft are also equipped with a controllable system for bringing in outside air to cool the cabin during hot weather.

Taxiing

Taxiing is the generally accepted word for manoeuvring an aircraft on either water or land surfaces. The prime purpose of taxiing is to manoeuvre the aircraft to the take-off position and return it to the apron after landing. Study the aerodrome chart and the runway and taxiway layout of all aerodromes you intend to use so that taxiing can be carried out expeditiously and safely.

Taxiways are identified by letters and spoken as Alpha, Bravo, Delta, Echo, etc.

Part of the taxi clearance that an aircraft could receive from a control tower might sound like this: ". . . the runway in use is two six — cleared to taxi via taxiways Bravo and Echo." This means that you first use taxiway Bravo, then turn onto taxiway Echo. You must know the taxiway and runway layout in order to determine the direction of turn from one taxiway to another. In the above taxi clearance the control tower has cleared your aircraft *to*, but not *onto* the runway in use; you will not have to ask permission to cross other runways. When the tower wants you to report crossing other runways it will be stated in your taxi clearance, but this does not relieve you of the responsibility of ensuring that other aircraft are not using these runways before you cross them.

Should you be told to "hold short" of *any* particular spot such as a taxiway, runway, or perhaps a terminal parking gate, that portion of the clearance must be "read back" to the controller.

At large, unfamiliar airports, especially those that use more than one runway simultaneously for take-offs and landings, taxiing an aircraft in the correct direction can be complicated and puzzling. Do not hesitate to ask the control tower or ground control for guidance if there is any doubt in your mind concerning correct procedure. Mention that you are a student pilot or unfamiliar with the airport; you will find Air Traffic Control personnel very co-operative and helpful.

Weathercocking. Taxiing a single-engine aircraft with a tail wheel in moderate to high winds can require effort and skill, due to its tendency to *weathercock* (continually wanting to head into wind). In the case of the aircraft with a nose wheel arrangement the weathercocking tendency is far less; as a result such aircraft are easier to control and manoeuvre on the ground, except under strong wind conditions.

Commence Taxiing. More engine power is required to start an aircraft moving, than to keep it moving. The amount depends on several things, but the principal factor is the degree of firmness of the surface upon which the aircraft is resting. In any case the throttle may have to be used more or less liberally, but once the aircraft starts to move, power must be reduced promptly. As soon as the aircraft starts moving, test the brakes by bringing it to a smooth full stop. The brakes may not operate perfectly, and it is important to know just how efficiently they are working.

Turning. Most light aircraft have steerable nose wheels or tail wheels connected to the rudder system, and under most circumstances they may be manoeuvred on the ground by this feature alone, without using brakes to assist turning. To turn the aircraft to the left, move the left rudder pedal forward; to maintain a straight heading, neutralize the rudder pedals.

The amount of rudder pedal action required to establish a desired heading varies according to the radius of turn, the condition of the manoeuvring surface, and the strength and direction of the wind. Of these three the wind has the most influence. In aircraft with differential braking systems (a separate brake system for each main wheel), a turn may be assisted by applying a sufficient amount of brake pressure on the same side as the rudder pedal being used to initiate the turn. Use brakes sparingly and never harshly.

45

Taxi speed and look-out.   It is considered poor practice to taxi an aircraft with excessive power settings and then control speed with continual braking. This may lead to overheating of the brakes and reduce their effectiveness. The lower the speed, the easier it will be to stop: taxi slowly and remember that there is a slight interval between the time the pilot initiates a change in speed or direction and the time the change actually occurs.

It is essential that the view ahead be unobstructed by the nose of the aircraft; in the case of many tail wheel aircraft, it may therefore be necessary to zigzag, first looking out one side of the aircraft and then the other, to ascertain that the taxiway is clear.

Use of controls while taxiing.   When taxiing directly cross-wind, if the control column is held to the side from which the wind is coming, the deflection of the ailerons will help maintain directional control. Various amounts of rudder will be required to prevent the aircraft from turning into the wind and to maintain a constant direction along the taxiway. When taxiing into wind and a right turn is desired, holding the control column to the left will assist the turn (and vice versa).

Take extra care when taxiing nose wheel aircraft, particularly those with a high wing, in strong *quartering tail winds*. To understand this term, picture an aircraft taxiing on a heading of, say, due north. Any wind blowing from a southeasterly or a southwesterly direction would be a quartering tail wind. When taxiing in strong quartering tailwinds, the elevators should be down (control column forward) and the aileron, on the side from which the wind is blowing, down (Fig. 2-6). As an added precaution, avoid sharp braking and sudden bursts of power.

You should realize that some manufacturers of nose wheel aircraft recommend the use of ailerons at variance to this method. In all cases the recommendations of the manufacturer should be followed.

When turning from downwind to upwind (into wind), slow down. The inherent tendency for an aircraft to turn into wind, further aggravated by too much forward speed, can introduce an excessive amount of centrifugal force. Coupled with top-heaviness, the additional centrifugal force may cause the aircraft to overturn. At the very least it may result in an undesirable side force being placed on various components.

When the manoeuvring area is soft or very rough, and more power is being used than would be normal for a firm level surface, it is best to taxi with the control column held firmly back, unless the wind strength and direction dictate otherwise.

Except when the air is still, the amount of foot pressure that must be applied to a rudder pedal to execute a manoeuvre or maintain a heading on the

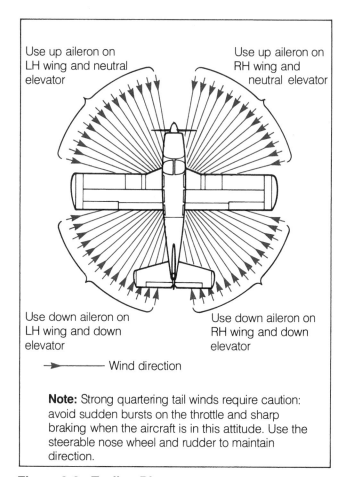

Use up aileron on LH wing and neutral elevator

Use up aileron on RH wing and neutral elevator

Use down aileron on LH wing and down elevator

Use down aileron on RH wing and down elevator

→ Wind direction

Note: Strong quartering tail winds require caution: avoid sudden bursts on the throttle and sharp braking when the aircraft is in this attitude. Use the steerable nose wheel and rudder to maintain direction.

Figure 2-6   Taxiing Diagram

ground depends basically on the wind velocity, the speed at which the aircraft is taxiing, and whether the aircraft has a tail wheel or a nose wheel landing gear arrangement. Of these three wind velocity is the most important.

Because of the many influencing factors, the aircraft's response to steering on the ground varies. In reasonably still air, steering can be almost as positive as that of an automobile, but control difficulties increase as wind velocity increases. Relatively little pressure on the rudder pedal is required to turn an aircraft into wind, but considerable pressure and coarse movement may be necessary to turn the aircraft out of wind. In addition, once the aircraft is started into the turn, this pressure may have to be varied to maintain the desired radius of turn. It may be necessary under some circumstances, especially in the case of a tail wheel aircraft, to anticipate the need for a control action before the requirement for such action becomes evident. For example, when turning into wind you may have to anticipate a need for opposite rudder pressure to arrest the turn, so that it does not become too sharp.

It is impossible to rule on a specific taxiing speed that will assure safety at all times. The primary requirement for safe taxiing is complete control of the aircraft, which means the ability to stop or turn where and when necessary. The speed should be slow enough that when the throttle is closed the aircraft may be stopped promptly. However, it is again emphasized that safe taxiing is directly related to wind velocity: the stronger the wind, the more slowly the aircraft should be manoeuvred on the ground.

When taxiing on a soft surface, such as a muddy field or in slush or snow, maintain speed at a slow, steady pace, otherwise the aircraft may come to a halt before power can be reapplied. This may necessitate the use of near full power to begin moving again. Besides being bad for the engine, this carries the risk of the propeller picking up lumps of mud, ice, snow, etc., damaging itself or other parts of the aircraft.

It is poor practice to turn or try to turn an aircraft by pivoting it about a stationary main wheel through the use of differential brake. If differential braking is used to assist a turn, allow the braked wheel to rotate forward sufficiently to avoid putting a twisting strain on the wheel and strut assembly.

When starting to taxi, first let the aircraft roll forward slowly and centre the nose or tail wheel. This will prevent the possibility of a swerve into another aircraft or a nearby obstruction.

Instrument Checks.   While taxiing in an area clear of obstacles and other aircraft, check the turn-and-bank indicator (T&B), attitude indicator (AI), and heading indicator (HI) for deflection, displacement, and indications as follows:

Instrument	Taxiing Straight	Turning Left	Turning Right
T&B Needle	Centre	Left	Right
T&B Ball	Centre	Right	Left
AI Bar	Steady	Steady	Steady
HI Degrees	Steady	Decreasing	Increasing

Marshalling.   When an aircraft is receiving outside guidance, it is being *marshalled*. It is always wise to have competent outside help when taxiing on an icy surface, in high winds, or in congested areas. The standard system of marshalling signals may be found in Transport Canada's *Aeronautical Information Publication* (A.I.P.) Canada.

Taxiway Courtesy.   Never block a taxiway unnecessarily. If for some reason the engine warm-up or run-up will cause delays to aircraft behind you, choose some other convenient place on the airport to carry out these functions.

Attitudes and Movements

In this exercise you will learn the range of attitudes through which the aircraft will normally be operated and how the movements necessary to achieve and maintain the desired attitudes of flight are produced and controlled. Some of these matters may appear complicated on paper, but you will gain understanding very quickly when they are demonstrated in the air.

Look-out

Now that flight training has begun in earnest, start observing this rule: *look around*. For safety in flight, *keep alert for other aircraft*. Look continually. Realize that there is a blind spot beneath your aircraft and never assume that others see you. Be especially alert during periods of nose-up attitudes of your aircraft, when the blind spot enlarges due to a decrease in forward visibility.

A pilot must be constantly on the look-out for other aircraft and must keep up a continuing search of the sky. It is commonly believed that the eye sees everything in its field with equal clarity. This is not so. Fix your gaze about 5 degrees to one side of this page, and you will no longer be able to read the printed material. Studies have revealed that the eye perceives very poorly when it is in motion. Wide sweeping eye excursions are almost futile and may be a hazard, since they give the impression that large areas of sky have been examined. A series of short, regularly spaced eye movements is recommended for maximum efficiency in searching the sky.

Transfer of Control

During flight training there must be a clear understanding, between the student and the flight instructor, of who has control of the aircraft at a given moment. Whoever is handing over control should say in clear tones, "You have control." This should be immediately acknowledged by the words, "I have control." When the flight instructor wishes to take over control of the aircraft, the instructor does so and at the same time says, "I have control." The student acknowledges immediately by saying, "You have control."

Attitudes

The basic attitude of an aircraft is termed a *cruise attitude*. Cruise attitude is the datum (reference point) to which all other attitudes of flight are related. It can be defined as the aircraft attitude for level flight at a constant altitude and airspeed, using a recommended cruise power setting, with the wings parallel to the horizon. All attitudes are considered as being relative to the horizon.

The attitudes of flight may be broken down into two groups:

Pitch Attitudes. Any attitude of the nose of the aircraft above or below the reference datum. Fig. 2-7 illustrates a range of pitch attitudes above and below the reference datum and indicates the approximate attitude limits for this stage of training. The attitudes above the datum are termed "nose-up attitudes," and those below, "nose-down attitudes."

Bank Attitudes. Any attitude of the wings of the aircraft when inclined relative to the datum. Fig. 2-8 illustrates a range of bank attitudes relative to the reference datum. The illustration indicates the approximate bank attitude limits for this stage of training.

The airspeed for flight in the cruise attitude selected should be noted, as it will be referred to when nose-up and nose-down attitudes are demonstrated. The other flight instruments will be referred to frequently. If you have an appreciation of the performance of your aircraft by reference to flight

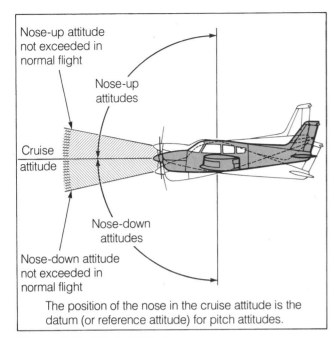

Figure 2-7 Pitch Attitudes

Nose-up attitude
not exceeded in
normal flight

Nose-up
attitudes

Cruise
attitude

Nose-down
attitudes

Nose-down attitude
not exceeded in
normal flight

The position of the nose in the cruise attitude is the
datum (or reference attitude) for pitch attitudes.

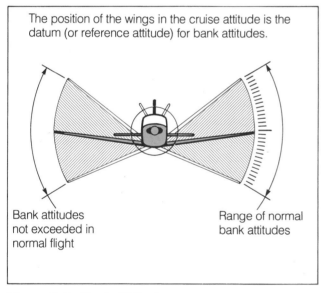

Figure 2-8 Bank Attitudes

The position of the wings in the cruise attitude is the
datum (or reference attitude) for bank attitudes.

Bank attitudes
not exceeded in
normal flight

Range of normal
bank attitudes

instruments, as well as by outside reference, you will develop from the start the habit of monitoring your own and the aircraft's performance continuously.

Handling the Controls

During the air exercises the position of the feet, when applying pressure to the rudder pedals, should be comfortable, with most of their weight supported by the heels in contact with the floor, thus allowing a fine sensitivity of touch with the toes. The control column should be held firmly but lightly with the fin-

gers, not grabbed and squeezed. Apprehension and tension may result in a tendency to choke the control column. Developing such a habit destroys "feel".

Control Response

The amount of control movement required to achieve a desired flight response depends to a great extent on the speed of the air flowing over or past the ailerons, elevators, and rudder. The cruise attitude airspeed may be considered the design datum for control effectiveness; at speeds above cruise speed, the controls become firmer, and there is a greater reaction to equivalent control movement. At lower cruising airspeeds, the controls become more yielding and less effort is needed to move them, but relatively more control movement may be needed to achieve an attitude change. The ailerons, being outside of the propeller slipstream, react consistently with airspeed changes, but the elevators and rudder do not (except in a power-off descent). Since the elevators and rudder are in the propeller slipstream, they will remain sensitive to control movement. This sensitivity increases with increase of power more or less independently of airspeed in the low ranges, until the cruise attitude airspeed is reached. In a power-assisted descent at a low airspeed, the aileron control will require relatively coarse movement, whereas elevator and rudder control movements will remain relatively fine to achieve the desired control response.

Movements, Controls, and Axes

The fundamental consideration is the method of control of the three movements of the aircraft. In any aircraft all three movements are around one central fulcrum, the Centre of Gravity, and they can be defined relative to the pilot and/or the aircraft as follows (Fig. 2-9):

1. **Pitching:** any movement around the lateral axis.
2. **Rolling:** any movement around the longitudinal axis.
3. **Yawing:** any movement around the normal (vertical) axis.

These movements are always relative to the aircraft and the pilot, regardless of the aircraft's position relative to the horizon. This may be difficult to visualize here but will become apparent in flight.

The three movements of an aircraft (Fig. 2-9), pitching, rolling, and yawing, are governed by the

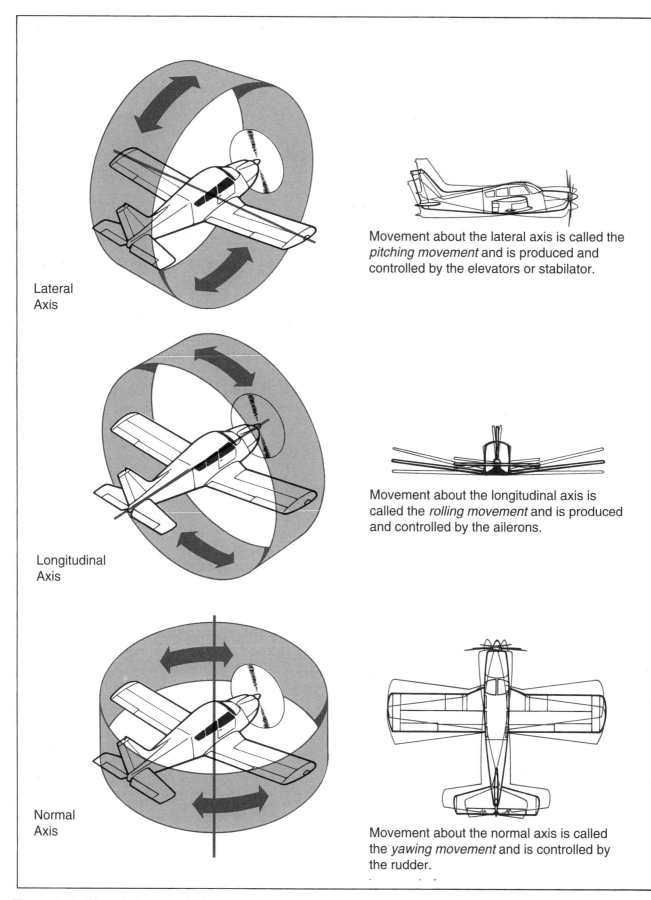

Lateral
Axis

Movement about the lateral axis is called the *pitching movement* and is produced and controlled by the elevators or stabilator.

Longitudinal
Axis

Movement about the longitudinal axis is called the *rolling movement* and is produced and controlled by the ailerons.

Normal
Axis

Movement about the normal axis is called the *yawing movement* and is controlled by the rudder.

Figure 2-9   Aircraft Axes and Movements

three controlling surfaces — elevators, ailerons, and rudder. The elevators are used to produce and control the pitching movement required to achieve and maintain the desired pitch attitudes. The ailerons are used to produce and control the rolling movement required to achieve and maintain lateral level and bank attitudes. The rudder is used to control yawing movement.

Yaw may occur adversely for many reasons: turbulence, power changes, or misuse of rudder. Failure to control yaw may cause the aircraft to slip or skid and ultimately roll; therefore, control of yaw is very necessary to maintain co-ordinated flight.

The following statements concerning the flight controls will be true regardless of the position of the aircraft relative to the earth.

1. When backward pressure is applied to the control column, the nose pitches up.
2. When forward pressure is applied to the control column, the nose pitches down.
3. When the aileron control is moved to the right, the aircraft rolls to the right.
4. When the aileron control is moved to the left, the aircraft rolls to the left.
5. When the left rudder pedal is pushed, the nose yaws to the left.
6. When the right rudder pedal is pushed, the nose yaws to the right.

Inertia

An aircraft possesses inertia. It tries to continue on its original path even when forces are introduced to change that path. Thus when the controls are moved there may be a slight lapse of time before the flight path changes, even after the attitude has been altered. The attitude of the aircraft is always referred to as relative to the horizon. The horizon referred to is the earth's natural horizon, but the inference also includes the horizon bar of the attitude indicator.

Flight Instrument Indications

During the pitch and bank demonstrations, the instructor will ask you to observe the indications of certain flight instruments.

When the aircraft is pitched into a nose-up attitude,

1. Airspeed decreases.
2. The miniature aircraft will be above the horizon bar of the attitude indicator.

When the aircraft is pitched into a nose-down attitude,

1. Airspeed increases.
2. The miniature aircraft will be below the horizon bar of the attitude indicator.

When the aircraft is rolled into a banked attitude for a turn,

1. The miniature aircraft will indicate a bank in relation to the horizon bar on the attitude indicator.
2. The heading indicator shows a change in direction.
3. The turn-and-bank indicator needle will be deflected in the direction of the turn; if the turn is co-ordinated, the ball will be centred.

When the aircraft is pitched up or down while banking left or right, the instrument indications will be a combination of those indicated during the individual attitude demonstrations. For example, should the aircraft be pitched into a nose-down attitude while in a co-ordinated bank to the left:

1. The airspeed will increase.
2. The miniature aircraft will be below and banked to the left in relation to the horizon bar of the attitude indicator.
3. The needle of the turn-and-bank indicator will be deflected to the left, and the ball will be centred.
4. The heading indicator will show decreasing degrees of heading.
5. The altimeter will show a constant decrease in altitude.

This is your first real training exercise. Many new and relatively strange events seem to be occurring rapidly. Even after the exercise has been demonstrated in the air, there are bound to be things you do not fully understand. This is perfectly natural, so do not hesitate to question your flight instructor concerning areas that still appear vague to you.

At a certain stage you may feel that you cannot do anything right, that you lack co-ordination and comprehension, that you aren't learning anything. All students go through this stage. Stay determined and you will discover that learning comprehension and co-ordination will sharpen as you become more familiar with the aircraft and the environment in which it operates.

Straight-and-Level Flight

Straight-and-level flight may be described as holding a steady direction with the wings laterally level while maintaining a constant altitude. This skill is essential in all of your subsequent training, so it is important to establish correct habits in this exercise.

Straight-and-level flight is achieved by the restrained use of all three flight controls. It is normal for minor variations to occur in heading, altitude, and airspeed. Pilots constantly strive for accuracy by checking the instruments and making prompt, small corrections to keep performance within close tolerances. In smooth air, the actual control movements are so small that it is more a question of applying slight pressure than any appreciable displacement of the flying controls.

In this exercise, straight-and-level flight will be demonstrated and practised, using the cruise attitude as the focal point around which variations of attitudes and airspeeds in straight-and-level flight will be achieved. The cruise attitude is established by visually fixing the relationship of some portion of the aircraft, usually the nose and the wing tips, with the horizon. As experience is gained, you will develop a sense of being level but the visual aids will be used as checks.

Straight flight is maintained by keeping the wings level and applying the necessary pressures on the rudder pedals to prevent yaw. If you allow the aircraft to bank, it will begin to turn in the direction of the lower wing.

Use the heading indicator to maintain straight flight. This instrument must first be set by readings taken from the magnetic compass and reset every fifteen minutes thereafter to remain accurate.

Magnetic Compass

Due to the construction of the magnetic compass, errors occur during flight in turbulent air or while turning or changing speed. The errors presented here, classified as acceleration error and turning error, are valid only for flight in the Northern Hemisphere.

As you accelerate on a heading of east or west, the compass indicates a turn to the north. The compass returns to its previous heading as the acceleration subsides. A deceleration will produce the opposite effect. The compass will indicate a turn to the south and will return to its previous heading as the deceleration diminishes. Acceleration/deceleration error is most pronounced on headings of east and west and diminishes to no effect on headings of north and south.

Turning error is most pronounced when you are turning from a heading of north or south. When you begin a turn from a heading of north, the compass initially indicates a turn in the opposite direction. Once the turn is established the compass will show the turn, but it will lag behind the actual heading. When you begin a turn from a heading of south the compass will indicate a turn in the correct direction, however, it will lead the actual heading. The amount of lag or lead diminishes as the turn progresses until the aircraft reaches a heading of east or west, at which point the turning error is nil.

When turning from a heading of east or west to a heading of south there is no error as you begin the turn. However, as the heading approaches south the compass increasingly leads the actual aircraft heading. The opposite is true when turning from a heading of east or west to a heading of north. In this case the compass increasingly lags the actual aircraft heading as it approaches north.

The magnetic compass reads correctly only when the aircraft is in straight unaccelerated flight. In other words, you can set the compass when flying straight in either level, climbing, or descending flight provided the airspeed is constant. When taking readings under turbulent conditions it may be necessary to take two or three readings and average the results or read the mean of the swing of the compass.

Effect of Power

The air pushed backward by a propeller revolves with considerable force around an aircraft in flight, in the same direction as the rotation of the propeller, causing an increased pressure on one side of the fin. This produces a force that moves the tail sideways and causes the aircraft to yaw. To compensate for this action, the fin may be slightly offset, either aerodynamically or mechanically, to provide optimum directional balance at normal cruise power settings. An increase in power, which increases the rotational force of the propeller slipstream, will cause aircraft with clockwise rotating propellers to yaw to the left. A decrease in power will cause the aircraft to yaw to the right. This action can be readily observed if power is increased or decreased. Any tendency for the aircraft to yaw with power changes should be anticipated and prevented by appropriate use of rudder.

When power is increased, the nose will pitch up and if no compensating control movement is made, the aircraft will begin to climb. Decrease power and the nose will pitch down and if no control adjustment is made the aircraft will start to descend. When power adjustments are made to increase or decrease airspeed in straight-and-level flight, immediate flight control adjustments are required to keep the nose from pitching up or down. As power is increased or decreased, keep the pitch attitude constant with appropriate elevator control pressure. When the aircraft is at the desired airspeed, trim to relieve the control pressure required to maintain straight-and-level flight.

Attitude Plus Power Equals Performance

A question often arises concerning which cockpit controls have primary control over air speed and altitude. The truth is that neither elevator nor throttle controls airspeed or altitude independently. The elevators control the pitch attitude of the aircraft, and the throttle controls the power. An expression you will hear many times in your flight career explains this: "Attitude plus power equals performance."

When an aircraft is in the cruise attitude, forward or backward pressure on the control column will affect both speed and height. Likewise, changes in power settings can affect both speed and height. This may prompt the question of which control takes precedence in a manoeuvre that requires adjustment of both power and flight controls. The fact is that the controls are interdependent, since changes in speed in the cruise attitude require simultaneous adjustment of the throttle and elevator controls. Any concept to the contrary will only delay progress in acquiring the control coordination needed to manoeuvre an aircraft from one airspeed to another.

Reducing Airspeed

To reduce the airspeed while in straight-and-level flight, throttle back smoothly to the power setting estimated for the speed desired. Anticipate and prevent any yaw that will occur as a result of the power change. Apply sufficient back pressure to the control column to maintain the desired altitude. Keep the wings level. At first, the back pressure needed will be barely perceptible, but as the speed decreases, the pressure needed will increase. When the desired airspeed is reached, readjust the power setting, then trim the aircraft.

Increasing Airspeed

To increase the airspeed while in straight-and-level flight, advance the throttle smoothly to the power setting estimated for the speed desired. At the same time, apply sufficient forward pressure to the control column to keep the altitude from increasing. Keep the wings level. When the desired airspeed is reached, readjust the power setting, then trim the aircraft.

Increasing and decreasing airspeed in level flight has important practical applications. For example, you need this skill to maintain correct spacing with other aircraft in the circuit. At a busy airport, aircraft have varying circuit and approach speeds. It may be necessary at one point in the circuit to maintain a high cruise speed and then reduce speed at another point to fit safely into the traffic pattern while maintaining a constant altitude.

Straight-and-level requires minimum use of the controls if the aircraft is properly trimmed and the air is smooth. In rough air, more physical effort and coordination are needed to maintain heading and altitude.

Trim

As you increase and decrease airspeed, you will feel pressure on the controls as you try to maintain level flight. The more the speed changes, the more pressure you will feel. Once the new speed is established, eliminate the control pressures by trimming the aircraft.

Trim is also influenced by a change in power. Pressure changes on the controls will be felt before there is any change in airspeed. The amount of pressure felt will depend on the amount of power change.

Changing attitude and power affects trim. In trimming the aircraft, it is not what you do that dictates whether you trim, but rather what pressure you feel on the controls.

Once your instructor shows you how to trim the aircraft, make trimming a habit. Learn to recognize control pressures and eliminate them to make your flying more accurate and less fatiguing.

Climbing

Climbing is the process whereby an aircraft gains altitude. During the first demonstration of this exercise, you will notice that there are changes in far more than altitude. When an aeroplane climbs, forward visibility decreases because of the higher nose attitude. Airspeed also changes and becomes an important reference. Maintaining a particular airspeed during the climb allows you to obtain optimum performance.

The question "What is optimum performance?" depends on the circumstances. There are a number of ways of climbing. If you are departing from an aerodrome where you must climb over an obstacle the best climb will be one that provides the steepest climb angle. If you depart on a long cross-country flight and wish to take advantage of strong tail winds at altitude, the best climb will be one that allows you to gain altitude as quickly as possible (Fig. 2-10). If you are climbing in the vicinity of other traffic, the best climb will be one with a pitch attitude low enough to afford good visibility. If you are just climbing with no particular concern or aim other than gaining altitude the best climb will be one that combines reasonable climb rate with good visibility and efficient engine cooling.

During your training, you will learn how to accomplish each of the following kinds of climbs:

1. Best rate
2. Best angle
3. Normal
4. En route

The airspeed at which you climb will determine which of the performances will be achieved.

Best Rate. The recommended best rate of climb speed is the airspeed that will afford the greatest gain in height in a given time. If it is important to reach a given altitude quickly, then this is the airspeed to use.

Best Angle. The best angle of climb is used to achieve the greatest gain in height in a given distance. If there are obstacles in the take-off path, for example, the aircraft should be climbed at the best angle of climb speed so that within the shortest possible ground distance the aircraft will be well above the height of the obstacles.

It is possible for the engine to become overheated if the aircraft is flown for too long a period at the best angle of climb or best rate of climb speed. The normal climb speed should therefore be resumed as soon as it is appropriate to do so.

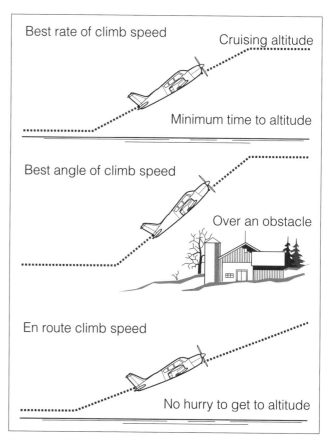

Figure 2-10   Climb Speed

54

Normal Climb. As the term implies, normal climb speed is the speed at which the aircraft is climbed under normal circumstances. Normal climb speed is higher than best rate of climb and best angle of climb speeds and is recommended for routine climbing situations because:

1. Forward visibility is better.
2. Most light aircraft take off and climb at full throttle. At this power setting the engine is dependent upon a high volume of airflow for cooling. Therefore the higher the airspeed the more effective the cooling. The normal climb speed specified for an aircraft takes into account and allows for (among other things) the need for adequate cooling.

En route Climb. En route climbs are carried out at various airspeeds between normal climb and normal cruise speed. The purpose could, for example, be to gain altitude slowly under a gradually upsloping cloud cover, using an airspeed slightly under normal cruise speed and a power setting slightly above cruise power. During this type of climb, convenience and comfort are the prime factors, since no climb time or climb distance limitations are assumed to apply.

As mentioned in Exercise 6, the propeller slipstream attempts to yaw the aircraft to the left. This effect becomes especially apparent at low airspeeds with high power settings, such as during a climb. Keep the wings level and use rudder to control any tendency for the aircraft to yaw to the left.

Before you begin a climb, look around carefully for other aircraft, particularly in the area ahead. As the nose of the aircraft rises, your forward scan becomes limited to the point where other aircraft at the same altitude or below become obscured.

To enter a normal climb, establish the aircraft in a nose-up attitude, one that you estimate will maintain normal climbing airspeed and increase power to the recommended setting for a normal climb. At first the operation of flight and power controls will be separately timed movements, but as you gain experience they should occur almost simultaneously. When the airspeed has settled, adjust the aircraft attitude to attain the desired airspeed. Recheck the power setting, then trim the aircraft until no pressure is required on the control column. The altimeter will show a steady increase in height, and the vertical speed indicator will show a steady rate of climb.

If the air is turbulent, even experienced pilots can have difficulty achieving steady indications on these instruments, especially the vertical speed indicator.

To return to straight-and-level flight from the climb, establish the aircraft in the normal cruise attitude, allow it to accelerate to cruising airspeed (care

must be taken not to exceed manufacturer's recommended RPM at this time), and reduce engine power for normal cruise flight. Readjust attitude to maintain altitude, recheck the power setting, then trim the aircraft for straight-and-level flight.

The density of the air plays an important part in the climb performance of an aircraft (Fig. 2-11). The more dense the air the better the performance. Three generally true factors to remember about air density are:

1. Density decreases as height increases.
2. Density decreases as temperature increases.
3. Density decreases as moisture in the air increases.

Humidity also affects aircraft engine performance because water vapour in the air reduces the amount of air available for combustion.

Relatively good climb performance can be expected on a cold, dry day from an aircraft at an airport with a field elevation of 150 feet above mean sea level, whereas a much poorer performance can be expected on a hot, humid day at an airport with a field elevation of 2,000 feet above mean sea level.

The air density, or lack of it, that affects the engine performance also affects the function of the airspeed indicator as the aircraft gains height. As the density of the air decreases with gain in height, the airspeed indicator indicates a progressively lower speed, although the actual speed of the aircraft may be relatively unaffected. Unless airspeed adjustments are made as the aircraft climbs, the rate of climb will decrease until finally the aircraft will be flying level instead of climbing.

To maintain a reasonably accurate rate of climb, a rule of thumb is to decrease the recommended indicated sea level climb speed by 1.75 percent (about 2 KT) for every 1,000 feet increase in height above mean sea level, excluding the first 1,000 feet. Suppose the recommended normal indicated climb speed of the aircraft is 90 KT. At an indicated altitude of 2,000 feet, the indicated climb speed would be adjusted to 88 KT; at 3,000 feet indicated, 86 KT; and so on.

Density altitude is the altitude corresponding to a given density in a standard atmosphere. It is a "condition," not a level of flight. Unless density altitude is known, it is difficult to determine the performance of an aircraft accurately, and this can be a very important factor under some conditions of take-off and climb. For example, the flight manual for a certain aircraft states that at an airport 3,000 feet above sea level the initial rate of climb will be 400 feet per minute. The elevation of the airstrip should be converted to density altitude to give a true picture of expected aircraft performance. Should the outside air temperature be

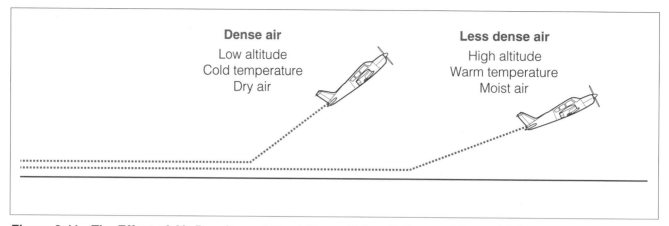

Figure 2-11   The Effect of Air Density and Humidity on Take-off Run and Rate of Climb

+28°C at the time, the density altitude could be as high as 5,000 feet. Looking at the flight manual again for an elevation of 5,000 feet, the initial rate of climb is reduced to 260 feet per minute. Some climb data tables give mean temperatures with field altitudes, but they must be interpolated carefully under extremes of temperature. Density altitude calculations can be resolved quickly on most circular slide-rule or electronic flight computers. As moist air (water vapour) weighs less than dry air, it is less dense. Therefore, on a moist or rainy day the resulting climb performance is less than on a normal dry day. Thus, under some conditions, the rate of climb after take-off could be critical.

An aircraft with retractable landing gear climbs at a higher rate, and gains height in a shorter distance when the landing gear is retracted. As soon as possible after the aircraft is established in its initial climb attitude, the landing gear should be retracted to obtain optimum climb performance. Be prepared to make attitude corrections as the landing gear retracts, since many aircraft are inclined to pitch upward when the wheels are no longer offering resistance to the -propeller slipstream.

On some types of aircraft a certain degree of flap is extended as a routine take-off procedure, but customarily flaps are used only when it is necessary to shorten the take-off run and steepen the initial angle of climb. In any case, consult the Aircraft Flight Manual for correct usage, since even the flap setting offering optimum lift for minimum drag will lower the aircraft's rate of climb. Therefore, if flaps have been extended for take-off, retract them as soon as there is no longer any operational requirement for them, but not before the aircraft is established in the desired climb attitude well above any obstacles.

Should there be a risk of carburettor ice, you may have to apply carburettor heat during the climb. In the case of full throttle climbs, this will reduce available engine power. Maintain airspeed and accept a slight reduction in rate of climb. In an aircraft designed to climb at power settings less than full throttle, when carburettor heat is applied, maintain the desired RPM (or manifold pressure) by advancing the throttle, and as air density allows, maintain airspeed and rate of climb. Consult the Aircraft Flight Manual for procedures in a specific aircraft.

The effect of weight on the performance of a light aircraft is more pronounced in a climb than any other normal manoeuvre. Because of ground effect, an aircraft may leave the ground within an acceptable distance when loaded to its maximum permissible gross weight, but once out of ground effect its rate of climb may be seriously affected. Climb performance data are of great importance if there are obstacles in the proposed climb path of the aircraft, especially if the airport (field) elevation is high and the ambient temperature is high. Consult the Aircraft Flight Manual for climb performance data.

Occasions arise when it is necessary to overshoot from an approach to landing and enter a climb when the aircraft is in the landing configuration with flaps fully extended. Take-off power must be applied smoothly but promptly as the aircraft is placed in a nose-up attitude consistent with a safe climb airspeed. With flaps fully extended, this attitude must be estimated by outside visual references based on a prior knowledge of the attitude. Since the trim was adjusted for a landing, the application of power will most likely pitch the nose upward; therefore, it is important that you be prepared to apply forward pressure on the control column and readjust the trim to maintain the desired attitude. The flaps must be retracted as soon as possible to an appropriate setting for the climb, but gradually and in small increments. Each time the flap setting is changed, the trim must be readjusted. For those aircraft with mechanically defined flap settings, retract the flaps one setting at a time.

During prolonged climbs, lower the nose momentarily, or change heading at regular intervals to search the sky ahead of the aircraft for other air traffic.

Descending

Descending from a higher to a lower altitude can be carried out in several ways to satisfy various operational requirements. Descents can be divided into two basic procedures:

1. Power-on (power assisted) descents.
2. Power-off (glide) descents.

Both basic methods of descending can be varied to meet the rate of descent and distance covered requirements of practically any situation. However, the power-on descent gives the pilot more control of the aircraft's descent path.

The airspeed that provides the best lift/drag ratio will permit the aeroplane to glide for maximum range. Determine and remember this airspeed, as this is the type of descent used for approaches to forced landings. Most Aircraft Flight Manuals include the best glide speed in their maximum glide charts. One typical chart shows that, at its best glide speed of 70 KT (in still air) from a height of 6,000 feet above ground level, the aircraft will glide a distance of 10 nautical miles (NM). The same aircraft flown at 60 KT may sustain a slightly lower rate of descent, but it will not attain the 10-mile distance, because of the lower airspeed. If the same aircraft is flown at an airspeed of 55 KT, the lift/drag ratio deteriorates to the point where the rate of descent is much greater than it would be at 70 KT. This together with the lower airspeed would achieve a glide distance of less than 4 miles.

The recommended best gliding speed will provide an attitude that achieves the greatest range (distance) in still air. However, wind velocity plays a commanding role in determining the airspeed and attitude to cover the greatest distance per unit of height available. Determining this attitude under the constantly varying conditions of wind direction and wind speed can be most difficult. So, in an emergency such as a forced landing, instead of complicating an already problem-atic situation, it might be better to use the familiar recommended still air gliding speed, then estimate the range available to you under the circumstances. A visual method for estimating gliding range is outlined in the following paragraphs.

Estimating Range

You will recall from a previous chapter that if another aircraft appears to occupy a stationary position on your windshield and to be growing larger, you will eventually collide with it unless evasive action is taken. In relative terms, the other aircraft becomes a stationary object and as it appears to grow larger you are on a collision course with it. This same principle may be effectively applied when attempting to reach a specific point on the ground during power-on or power-off descents, such as when executing a normal approach, a precautionary landing, or a forced landing. The point on the ground may be the point of flare at an aerodrome or the touchdown point of a forced landing. If the selected spot on the ground remains stationary in relation to a fixed point on your windshield, the aircraft will subsequently touch down at the selected spot on the ground.

The fixed point is any point on the windshield that you choose as a reference point. It could be so many inches up from the instrument panel, or adjacent to the magnetic compass, a mark you have made yourself with a non-permanent marking pencil, or a small piece of tape. Everything is measured in relation to this imaginary, or actual, point on the windshield.

When stabilized in a constant power or power-off descent at a constant attitude and airspeed, visual observations of ground positions in relation to a fixed point on your windshield will provide information as follows:

1. Positions on the ground that appear to move down

from the fixed position on the windshield are ground positions that you can reach and fly over with height to spare.

2. The position on the ground that remains stationary in relation to the fixed position on your windshield is the ground position that your aircraft should reach.

3. Positions on the ground that appear to move up from the fixed position on the windshield are ground positions that your aircraft cannot reach.

Power-off Descents

To enter a power-off descent from straight-and-level flight:

1. Complete any cockpit checks and note the altimeter reading.
2. Search the sky, above and below, for other aircraft.
3. Close the throttle smoothly but promptly.
4. Keep straight (the aircraft will tend to yaw to the right) and allow airspeed to decrease.
5. Assume the approximate attitude for best glide airspeed.
6. Trim.
7. If necessary make minor pitch adjustments to attain correct airspeed, and retrim.
8. Note the steady decrease in altitude on the altimeter and the rate of descent on the vertical speed indicator.

To return to straight-and-level flight from a power-off descent:

1. Search the sky ahead, and above, for other aircraft.
2. Note the altimeter reading.
3. a) Advance the throttle to the power setting for cruise flight (carburettor heat off).
   b) Assume the cruise attitude and maintain it until the aircraft accelerates to cruise speed.
   c) Keep straight (the aircraft will tend to yaw to the left as the throttle is advanced).
4. Trim.
5. Adjust power and flight controls to maintain the desired airspeed and altitude.
6. Retrim.

During power-off descents the engine must not be allowed to become too cool, otherwise it may fail to respond properly when the throttle is advanced to regain straight-and-level flight. Cruising power should be applied at appropriate intervals during the descent to keep engine temperatures near normal and to prevent fouling of the spark plugs. Many aircraft require carburettor heat during power-off descents; in others it is not recommended. Consult the Aircraft Flight Manual for the correct procedure.

Power-on Descents

A power-on descent is used when precise control of the rate of descent and distance attained is desired. Most routine descents and approaches to landings are power assisted to control the rate of descent for passenger comfort and meet the speed and spacing demands of airport circuit procedures. To enter a power-on descent, carry out the cockpit checks, look around for other aircraft, then:

1. Reduce engine power to an RPM setting judged (or predetermined) to give a desired airspeed and rate of descent.
2. Allow the airspeed to decrease to that desired airspeed.
3. Lower the nose to an attitude that will give the desired rate of descent.
4. Trim to maintain this attitude.
5. Check that the airspeed and rate of descent are those desired; if not, increase or decrease the amount of power, adjusting the nose attitude as required, until the proper flight condition is obtained.
6. Retrim.

To decrease the rate of descent while in a power-on descent, apply the amount of engine power that will give the desired rate of descent, and simultaneously adjust the attitude of the aircraft to maintain the best descent speed and retrim for the new attitude. In most aircraft the application of power will pitch the nose up, and as a result very little control adjustment is needed to establish the new attitude and maintain the original airspeed. If the need for a power-off descent ensues, as the throttle is closed the nose will pitch down and more or less assume the attitude for best gliding speed without a great deal of adjustment, if the aircraft is properly trimmed for each attitude. Proper trim is the key to a smooth and accurate transition from one attitude or airspeed to another.

Any variation of airspeed and rate of descent may be combined to obtain the effect desired. An en route descent is usually a power reduction to provide a suitable rate of descent while still maintaining cruising airspeed. When approaching the destination airport, it may be desirable to reduce power so that descent is made at a reduced airspeed. Finally, the power-on descent involves a power setting that will provide the desired rate of descent while maintaining the recommended approach to landing airspeed.

Power-on descents requiring steep angles of descent, as in an obstacle clearance approach, usually involve full extension of the flaps and lower than normal approach airspeeds. Consult the Aircraft Flight Manual or Pilot Operating Handbook to determine the flap setting and airspeed. Should "calibrated airspeed" (CAS) be given, ensure that you use the airspeed correction table in the manual or handbook to calculate the "indicated airspeed" (IAS) to fly the obstacle clearance approach.

The angle of descent of aircraft with retractable landing gear can be effectively steepened by extending the landing gear. Most aircraft have too flat a glide angle for many purposes, such as clearing obstacles on the approach to landing. The angle of descent could be steepened by descending at a very high or a very low airspeed, but neither is satisfactory under normal circumstances. Therefore the aircraft is fitted with flaps. Flaps, when extended, steepen the angle of descent for any particular airspeed (Fig. 2-12). The more the flaps are extended, the steeper the angle of descent for a given airspeed. Another advantage is that the steeper the descent attitude, the better the view ahead. In addition to the drag caused by extending the flaps, lift is increased for the same airspeed. It is therefore possible to descend at a lower airspeed than when the flaps are retracted.

The prevailing wind will affect the maximum gliding distance greatly (Fig. 2-13). When gliding into a head wind you can increase range by selecting an airspeed slightly higher than normal. When gliding downwind a greater distance can be covered by using an airspeed lower than normal. The same logic applies when cruise flying for range into, or with, a wind.

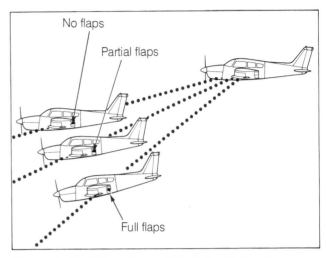

Figure 2-12 Wing Flaps Affect the Rate of Descent and Range

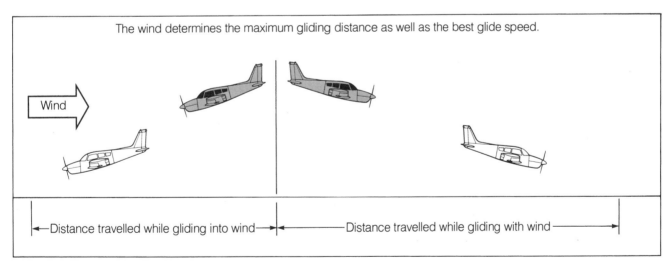

Figure 2-13 Gliding Distance and Gliding Speed

Turns

The *turn* is a basic manoeuvre used to change the heading of an aircraft. An accurate level turn may be described as a change of direction, maintaining a desired angle of bank, with no slip or skid, while maintaining a desired altitude. This is also the description of a co-ordinated turn. Aerodynamically the turn is probably the most complex of basic manoeuvres and involves close co-ordination of all controls.

To turn an aircraft, co-ordinated pressures are applied to the controls until a desired banked attitude is achieved. The object of bank during a turn is to incline the lift so that in addition to supporting the aircraft, it can provide the necessary force (centripetal force) toward the centre of the turn to oppose centrifugal force, which is endeavouring to pull the aircraft away from the centre of the turn.

In a level turn, lift must be sufficient both to support the aircraft and to provide the inward force. Therefore, it must be greater than during straight-and-level flight. The additional lift can be acquired by increasing the angle of attack of the wings and accepting a varying degree of reduction in airspeed (Fig. 2-14). Up to a certain degree of bank, airspeed may be maintained by increasing power.

For training purposes turns are divided into three classes:

1. **Gentle** turns, involving angles of bank up to 15 degrees.
2. **Medium** turns, involving angles of bank from 15 to 30 degrees.
3. **Steep** turns, involving angles of bank over 30 degrees.

In addition to level turns there are:

1. **Climbing** turns, which are normally gentle turns.
2. **Descending** turns, which may be gentle, medium, or steep.

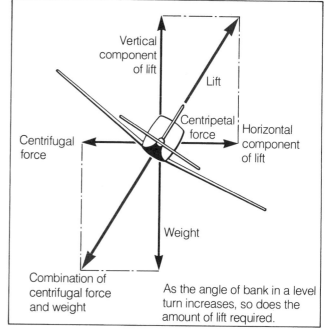

Figure 2-14 Forces in a Turn

Entering a turn, the rising wing creates more drag than the descending wing. This tends to yaw the aircraft toward the raised wing and causes the aircraft to attempt initially to turn in the wrong direction. This is *adverse yaw* (see Chapter 1). To counteract adverse yaw, use appropriate rudder pressure, in the direction of the turn. The amount of rudder movement necessary varies according to the abruptness of the execution of the turn, and often the type of aircraft.

It is most important at this stage to understand fully that rudder is used in a turn only if there is any adverse yaw. Adverse yaw has almost been eliminated from aircraft of recent manufacture, so that the rudder pressure required on turn entry to counteract adverse yaw has been reduced very effectively.

These are some basic facts you should understand:

1. At a given airspeed, the greater the angle of bank:

(a) The greater the rate of turn.
(b) The smaller the radius of turn.
(c) The higher the stalling speed.
(d) The greater the load factor.
2. The higher the airspeed at a given angle of bank:
   (a) The lower the rate of turn.
   (b) The larger the radius of turn.
3. To achieve a turn of the smallest radius and greatest rate for a given angle of bank, fly at the lowest possible airspeed for that angle of bank. A fact applicable to all aircraft is that the stalling speed increases as bank angle increases, and this increase in stalling speed accelerates very rapidly as the angle of bank continues to steepen. At 30 degrees of bank, the stalling speed increases to about 8 percent over the stalling speed in level flight; at 45 degrees, 18 percent; at 60 degrees, 40 percent; at 75 degrees, 100 percent; at 83 degrees, 200 percent. Therefore, an aircraft with a level flight stalling speed of 50 KT has a stalling speed of 150 KT (50 + 100) when subjected to an 83 degree bank turn.

Similarly, the wing loading or load factor increases slowly at first as the angle of bank increases and very rapidly thereafter (see Chapter 1, "Load Factor"). As an example, the load factor on an aircraft executing a 60 degree bank turn at any flying speed while maintaining altitude is double that of straight-and-level flight, while the load factor for an 80 degree bank turn is 5.76.

The importance of look-out, or searching the sky for other aircraft before and during a turn, cannot be overemphasized. Before entering a turn, look around carefully in both directions, above and below. During the turn continue to look out, especially in the direction of the turn. When recovering from the turn look around again, in all directions, above and below. To maintain a good look-out and manage the aircraft at the same time requires the pilot's constant attention.

Posture is important in all aircraft manoeuvres, but especially so in turns. Sit comfortably upright; do not lean away from the centre of the turn, but do not make a conscious effort to keep your body stiffly vertical. Relax and ride with the turn. Stiffening up or continually changing sitting position affects visual references and may cause handling of the controls to become tense and erratic (Fig. 2-15).

An accurate level turn entry requires that the aircraft be flying straight-and-level, as accurately as possible, prior to entering the turn. Any error made before entering the turn is likely to be exaggerated as the turn develops. The same principle applies to climbing and descending turns.

When entering, maintaining, or recovering from a turn, the trained pilot co-ordinates the movement of

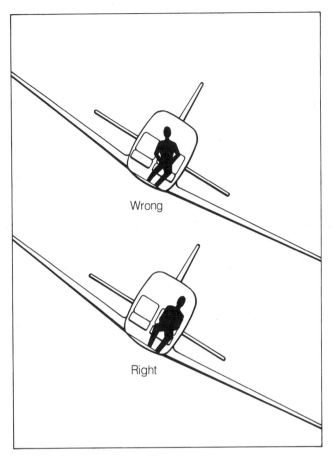

Figure 2-15 Posture During Turns

all three controls into one simultaneous movement. Initially, however, it is best to think of each control as having one definite function, the ailerons controlling bank attitude, the elevators controlling pitch attitude, and the rudder controlling yaw. This will help eliminate some of the more common faults at the outset, such as overcontrol of the rudder.

To enter a level turn:

1. Make sure that the aircraft is in accurate straight-and-level flight.
2. Look around for other aircraft.
3. Roll the aircraft gently to the desired bank attitude with aileron control. Maintain this attitude.
4. At the same time, use appropriate rudder pressure to control any tendency for the aircraft to yaw adversely.
5. Use elevators to maintain the aircraft in the correct pitch attitude in relation to the horizon.
6. Maintain the look-out.

In a gentle turn, the position of the nose in relation to the horizon, which is the visual reference for pitch attitude, will remain almost the same as in straight-and-level flight. However, as the angle of bank is increased beyond a gentle turn, the pitch attitude must be altered (by backward pressure on the control column) to increase lift; this is to compensate for the added load factor imposed by centrifugal force as the turn steepens. The loss in airspeed, or the need for an increase in power to maintain airspeed, becomes more apparent as the angle of bank increases.

As the aircraft settles into an accurate turn:

1. The nose will move steadily around the horizon, neither rising nor falling.
2. The airspeed will be constant.
3. The turn indicator will show a constant rate of turn.
4. The ball will be centred in its glass tube.
5. The altimeter will be steady on the selected altitude.

One of the most common faults when entering turns is excessive use of rudder. This fault can be corrected quickly or completely prevented if you remember right from the outset not to apply rudder unless it is necessary to control adverse yaw.

To recover from a turn:

1. Look around.
2. Roll the wings level with aileron control.
3. At the same time, use appropriate rudder pressure to control adverse yaw.
4. Keep wings level.
5. Maintain correct pitch attitude with elevator control.
6. Keep straight.
7. Look around.
8. Trim.

In gentle level turns the lateral stability designed into the aircraft will attempt to return it to straight-and-level flight, therefore, slight aileron pressure may be required to maintain it in such a turn. However, as the bank attitude increases beyond a certain angle, lateral stability is overcome. Simply stated, this results from the outer wing travelling faster than the inner wing and therefore obtaining more lift, causing the aircraft to continue its roll unless the pilot takes some action to stop it. Therefore, aileron control must be used accordingly to maintain the desired angle of bank.

Climbing and descending turns are executed like level turns except that, instead of maintaining a constant altitude, a constant climb or descent is maintained. While the control inputs to enter, maintain, and recover from the turn are the same as in level turns, there are additional considerations regarding power and attitude control.

In a climbing turn, additional power is required to achieve the desired increase in altitude. The nose-up attitude selected and maintained during the climbing turn will depend upon the operational requirements at the moment and will be those attitudes described in Exercise 7, "Climbing." In a descending turn, power will be reduced in varying amounts from cruising to throttle fully closed. The nose-down attitude will again vary to achieve the desired results, but will correspond to those attitudes as discussed in Exercise 8, "Descending."

The lateral stability of an aircraft in a climbing or descending turn is affected by the angle at which the relative airflow meets each wing. As a result:

1. In a descending turn, the aircraft moves a given distance downward during a complete turn, but the inner wing, turning on a smaller radius, descends on a steeper spiral than the outer wing, like the handrails on a spiral staircase. Therefore, the relative airflow meets the inner wing at a greater angle of attack and so obtains more lift than the outer wing. The extra lift acquired this way compensates for the extra lift obtained by the outer wing due to its travelling faster. Therefore, in a power-off descent the angle of bank will tend to remain constant.
2. In a climbing turn, the inner wing still describes a steeper spiral, but this time it is an upward spiral, so the relative airflow meets the inner wing at a smaller angle of attack than the outer wing. In this case the outer wing obtains extra lift, both from its extra speed and its greater angle of attack. Therefore, the angle of bank will tend to increase and the aileron control must be used accordingly to maintain the desired angle of bank.

Initially, climbing and descending turns will be entered from normal straight climbs and descents and the recovery made back to straight climbing or descending flight, to enable you to experience and readily observe the difference in pitch attitude necessary to maintain the desired airspeed. As you gain proficiency, these turns will be entered directly from straight-and-level flight and recovery made directly back to straight-and-level flight.

Power-off descending turns are particularly important, as they are directly related to forced landing procedures. It is necessary that you learn to execute this type of turn to reasonable proficiency almost subconsciously, since during a forced landing you must attend to many other details. Because the controls may be less responsive than in power-on turns, power-off descending turns require the development of a different technique from that required for power manoeuvres.

When recovering from a power-off descending turn, the pressure exerted on the elevator control during the turn must be decreased, or the aircraft will pitch up too high and airspeed will be lost. Such an error will require a lot of attention and control adjustment before the correct attitude and airspeed can be resumed. More altitude will be lost in a power-off descending turn than in a straight descent.

Steep Turns

Steep turns are a means of turning quickly in a relatively small area, but as an exercise in flying they have a value beyond purely practical application. They provide one of the best instances of sustained extra loading effect, together with excellent practice in co-ordinating all three flight controls and the power control. The practical applications of steep turns are almost all limited to emergency situations. Practise them diligently so you can execute them accurately and without hesitation.

Up to a limited angle of bank, a steep turn may be executed without increasing engine power. However, in order to maintain a constant altitude, airspeed must be sacrificed. When carrying out this procedure remember that the stalling speed increases with an increase in angle of bank (Fig. 2-16). (In a 60 degree

bank, an aircraft with a level flight stalling speed of 50 KT would stall at 70 KT.)

The greater the angle of bank, the greater the amount of lift required to maintain a constant altitude. Increased lift produces increased drag, thus more engine power is required to maintain a constant airspeed. Therefore, the angle of bank that can be sustained in a level turn (disregarding structural limitations) depends on the engine power available (Fig. 2-17).

A steep turn is entered like any other turn, but as the angle of bank is increased beyond the 30 degree angle of the medium turn, you will need extra engine power to maintain altitude and airspeed. A steep turn requires complete and simultaneous co-ordination of all controls. Because of the rapid rate of change of direction, the look-out for other aircraft before doing a steep turn is even more important than for other turns.

Enter as for a medium turn. As the bank increases beyond that of a medium turn, move the control column back to maintain the correct pitch attitude. Start increasing power as necessary to maintain airspeed as the angle of bank increases beyond 30 degrees. When the required angle of bank is reached, keep it constant with aileron control.

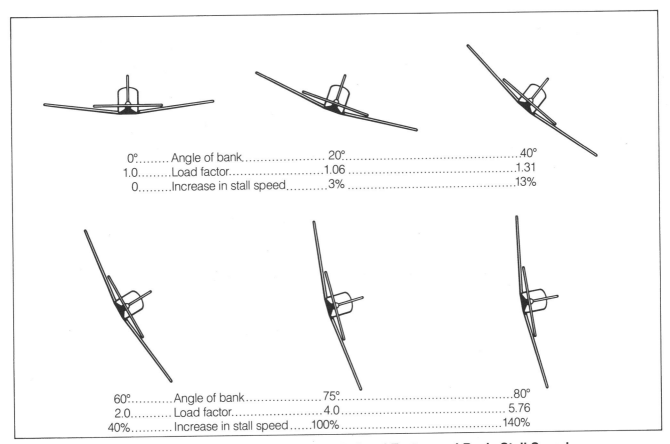

Figure 2-16   The Relationship Between Angle of Bank, Load Factor, and Basic Stall Speed

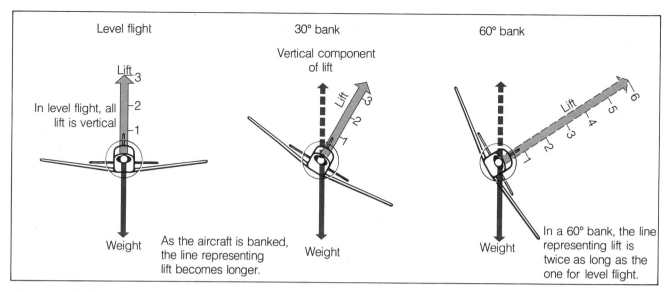

Figure 2-17   Lift and Angle of Bank

If the nose pitches too far down in a steep turn, do not attempt to correct by applying back pressure alone, as this may serve to tighten the turn. Use co-ordinated aileron and rudder pressure to reduce the angle of bank slightly and correct the pitch attitude.

Recover from the turn exactly as from any other turn, except that engine power should be reduced simultaneously with the return to straight-and-level flight. Maintain a good look-out.

As it gives you the ability to turn quickly in a reduced radius, the steep turn has important practical applications. For example, by increasing the rate and reducing the radius of turn, a steep turn can be used to avoid a collision. The reduced radius also makes a steep turn useful as a "canyon turn." This type of turn may require both a steep angle of bank and a reduced airspeed. The suggested airspeed to use is not less than the airspeed for maximum endurance. You will recall that maximum endurance airspeed is some-where between the airspeed for slow flight and the airspeed for maximum range. Extending a small amount of flap in most aircraft will reduce the stall speed, thereby providing an increased margin of safety. It is seldom necessary in practice or in an emergency to carry this turn beyond the 180 degrees needed to reverse direction in a canyon.

After you have performed the look-out, and the air-craft is at the desired airspeed in straight-and-level flight, enter the turn promptly with co-ordinated use of all controls. As the aircraft approaches a 30 degree angle of bank, apply power and establish the required bank. Maintain the bank and adjust the pitch attitude as necessary to control altitude.

When recovering from the turn, establish straight-and-level flight and allow the airspeed to increase before raising the flaps and reducing power.

A steep descending turn can be used to come down through a hole in clouds should you be unable to maintain adequate reference to the ground or water. Care should be taken in a steep descending turn to maintain a safe and constant airspeed and avoid a spiral.

Specific technique in these applications of steep turns — bank angle, power, airspeed, and flap settings — may vary according to the situation and type of aircraft. Your instructor will discuss these with you.

It is important to note that many of the situations that require the use of a steep turn are a result of poor decision making. Superior decision making can keep you out of situations requiring the use of superior skill.

Instrument Indications

During a turn instrument indications are as follows:

Turn-and-Bank Indicator.   The needle will deflect in the direction of the turn and will indicate the rate at which the aircraft is turning. In a co-ordinated turn, the ball will be centred in its curved glass tube. If the ball is off-centre to the inside of the turn, the aircraft is slipping into the centre of the turn. If the ball is off-centre to the outside of the turn, the aircraft is skidding out from the turn.

Attitude Indicator.   The horizon bar (of most instru-ments) will remain parallel to the real horizon, and the miniature aircraft, in relation to the horizon bar, will indicate a bank in the same direction as the real aircraft. This instrument also indicates the attitude of the aircraft in the pitching plane. The nose of the miniature aircraft, in relation to its artificial horizon, corresponds to the pitch attitude of the nose of the real aircraft in relation to the real horizon.

Heading Indicator.   Immediately as a turn begins, this instrument begins rotating to indicate the successive new headings of the aircraft during the turn. When the turn stops, it stops. To decrease the numerical values on the face of the instrument turn left; to increase values turn right. A memory aid is "left for less."

Airspeed Indicator.   Because the load factor increases as a result of the turn, additional lift must be obtained by increasing the angle of attack. This in turn creates more drag, resulting in a decreased airspeed. In a steep turn the airspeed decrease is more noticeable than in a gentle or medium turn. In the case of a poorly co-ordinated turn, the airspeed indicator will react more significantly. As well, if the nose is allowed to pitch too high, there will be a decrease in airspeed; conversely, if the nose is allowed to pitch too low, the airspeed indicator will rapidly indicate an increase in speed.

Altimeter.   In a co-ordinated level turn, the altimeter needle would remain stationary at the selected altitude. If the nose is held too high, there will be an increase in altitude. If the nose is allowed to drop too low, a decrease in altitude will be indicated.

Flight for Range and Endurance

To make effective use of an aircraft, you must understand the concepts of range and endurance and be able to use available charts to determine range and endurance in known circumstances.

Flight For Range

When an aircraft is being flown for maximum range, the objective is to fly the greatest distance possible per unit of fuel consumed. This capability depends on a number of factors.

Fuel Available

Perhaps the most obvious factor affecting range is the amount of fuel available. When the tanks are full, it is an easy matter to determine the range for the aircraft by simply referring to the appropriate charts, which normally assume full fuel, in the Aircraft Flight Manual.

However, there are occasions when it is not possible to carry full fuel, particularly when aircraft gross weight is a consideration. In these circumstances, it is simply a matter of determining the rate of fuel consumption for a given power setting and dividing the amount of fuel carried on board by that figure.

Angle of Attack

As far as the aerodynamics of the aircraft are concerned, maximum range is achieved when the aircraft is being operated at the angle of attack giving the greatest ratio of lift to drag. The angle of attack that gives the best lift/drag ratio for a given aircraft will always be the same and is not affected by changes in altitude or gross weight.

However, since most light aircraft do not have an angle of attack indicator, the pilot must rely on some other means to determine when the aircraft is being flown at the correct angle of attack.

Airspeed

There is an indirect relationship between airspeed and angle of attack. By referring to the chart in Fig. 2-18, you can see that flying at the correct angle of attack results in minimum drag, as shown on the total drag curve, and corresponds to a given calibrated airspeed. You can use indicated airspeed since the difference between the two speeds in most light aircraft is negligible. This is fortunate since it is the only speed to which you have direct reference.

So, there is an indicated airspeed that corresponds to the angle of attack that provides the best lift/drag ratio. This speed does not change with altitude, but must be increased slightly for increases in aircraft gross weight.

Aircraft Weight

Increasing the weight carried by an aircraft results in an equal increase in the amount of lift required to maintain level flight. This can be accomplished by either increasing the airspeed or the angle of attack. Since there is only one angle of attack that will produce the best lift/drag ratio, the only way we can generate the extra lift required is to increase the airspeed. As more power will be required to increase the airspeed, more fuel will be used per mile flown, and range will be decreased.

Note that manufacturers of light aircraft normally base performance data for their aircraft on gross weight. This doesn't mean that you should decrease your speed when operating at weights below gross.

66

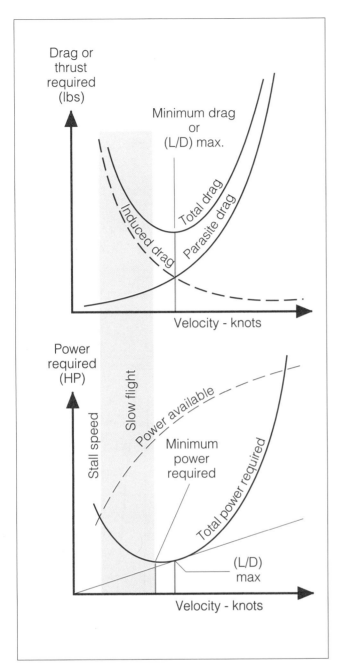

Figure 2-18 Power/Drag Curves

Although technically correct, this would not have a significant effect on range performance for most light aircraft. Besides, manuals for these aircraft don't give range performance charts for aircraft weights other than gross weight.

Centre of Gravity

One of the ways pitch stability is achieved in aircraft design is to have a positive force exerted downward on the stabilizer. This is usually accomplished by fixing the stabilizer to the aircraft in a manner that causes

the airflow to exert this force. This force is equivalent to adding weight to the tail and, depending on the type of aircraft, could be several hundred pounds. This is the same as carrying another passenger with baggage insofar as it affects the aircraft's performance.

When the Centre of Gravity is at its forward limit, more down force is required on the tail. This is accomplished by trimming the elevator to a slight nose-up position, which produces the following undesirable results:

1. More lift will be required (see the previous section on Aircraft Weight).
2. The position of the trim tab and the elevator will cause an increase in drag.

Since the solution to both of these problems is an increase in power to maintain the correct airspeed, our range will be reduced.

Conversely, with a Centre of Gravity at the aft limit of the Centre of Gravity envelope, we will increase our range because there will be less down force on the stabilizer and less drag caused by the elevator trim. Therefore, less power will be required to maintain the correct airspeed for range.

So, when considering Centre of Gravity and its effect on the range performance of an aircraft, it is best to have an aft Centre of Gravity.

Altitude

Selection of an altitude to fly is based on many factors, including wind, turbulence, ceiling, distance to fly, terrain, radio reception, map reading, and aircraft performance. Altitude has a significant effect on the range performance of an aircraft. The best altitude at which to fly is determined by the efficiency of the engine (and to some extent by propeller efficiency) and not by the aerodynamics of the aircraft design.

Engine Efficiency

At low altitude, the engine power output is controlled with the venturi in the carburettor partially closed by the throttle valve to avoid exceeding the recommended power settings. Restricting the flow of the fuel/air mixture to the engine in this manner causes it to run less efficiently than when the throttle valve is fully open.

To use both the aircraft and engine to best advantage to obtain maximum range, you must choose a higher rather than a lower altitude at which to fly.

The optimum altitude is the altitude that permits the throttle to be fully open while providing the power necessary to fly at the correct airspeed or angle of attack. At altitudes above this, more power would be required to maintain the correct airspeed. This will not be possible as the engine is already operating at full throttle.

Climb

In many cases, it may not be practical to climb to the optimum altitude for maximum range due to the nature of the intended flight, weather conditions, and the fuel required to climb to that altitude.

Referring to the chart in Fig. 2-19, you can see that range decreases for the 45 percent power setting from sea level up to 12,000 feet. This appears to contradict what was said about selecting a higher altitude for maximum range. However, a glance at the notes provided in the chart reveals the reason. Allowances have been made for the fuel used for engine start, taxi, take-off and climb, the distance travelled during the climb to altitude, and 45 minutes reserve fuel. These allowances reduce the amount of fuel available for achieving maximum range at the optimum altitude.

Wind

One of the conditions specified for the chart shown in Fig. 2-19, is that these range figures are based on a no wind situation. Depending on its strength and direction, wind may have a greater effect on the range of an aircraft than any other factor and is an important consideration when selecting an altitude at which to fly.

As wind speed generally increases with altitude, it may be prudent to select an altitude where any head-wind component will have minimal effect, and tail-wind components will have maximum effect. How does a head-wind affect range? Quite simply, it reduces your ground speed and increases the length of time to get to your destination. This results in more fuel being used.

You will never get the same range with a head-wind component as you would in calm wind or with a tail wind. However, with a head wind, you may improve the situation slightly by increasing the indicated air-speed required for range by approximately 5 to 10 percent (a negligible increase for most light aircraft) allowing you to get to your destination a little sooner. Consequently, by arriving sooner you have decreased the time you have been affected by the head wind without appreciably increasing fuel consumption.

As an exaggerated illustration, consider a typical light aircraft with fuel for 5.5 hours of flight at a rec-ommended range cruising speed of 80 KT, planning

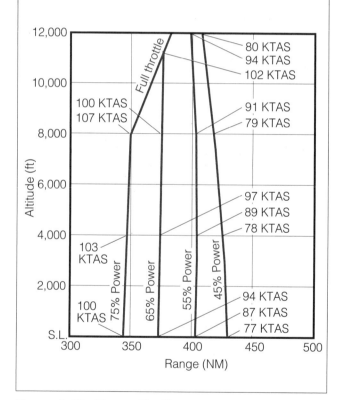

Range Profile

45 minutes reserve
24.5 gallons usable fuel

Conditions:
1670 lbs
Recommended lean mixture for cruise
Standard temperature
Zero wind

Notes:

1. This chart allows for the fuel used for engine start, taxi, takeoff and climb, and the distance during climb.

2. Reserve fuel is based on 45 minutes at 45% BHP and is 2.8 gallons.

Figure 2-19   Range Profile

to fly to a destination 170 miles away in a 50 KT head wind. (For illustration purposes, disregard the reserve fuel requirement.) At a ground speed of 30 KT and fuel for 5.5 hours, the aircraft would run out of fuel at 165 miles, leaving it 5 miles short of its destination.

The same aircraft with power increased to give it a higher cruising speed of 90 KT has also increased its fuel consumption so that only 4.5 hours of fuel is available, but at a ground speed of 40 KT the aircraft will cover 180 miles on its available fuel, 10 miles more than required to reach its destination.

Conversely, by reducing the indicated airspeed by the same amount (5 percent to 10 percent) with a tail wind, you can increase the benefit derived from this situation due to the increased ground speed and a slight decrease in fuel consumption that will result in an increase in range.

Determining Range

When we speak of range, we are simply talking about the distance the aircraft will travel in numerous highly variable circumstances. The cruise performance and range charts provided by the aircraft manufacturer make it possible to achieve optimum performance from your aircraft for a given set of these circumstances and to determine the range for a particular power setting. You must be proficient in the use of these charts to determine range information for your flights.

Using the Charts

First look carefully at the conditions specified for the Range Profile and Cruise Performance charts. For the Range Profile chart (Fig. 2-19), you will note that the figures are based on standard temperature, zero wind, and a weight of 1,670 pounds, which happens to be the gross weight for this aircraft. Another important point to note in the conditions specified for both charts is that the mixture must be leaned in accordance with the manufacturer's recommended procedures. Again, read the applicable conditions carefully. They are not the same for all charts.

When considering aircraft performance, true airspeed and fuel consumption for a given power setting at a given altitude are obtained from a chart in the Aircraft Flight Manual such as the one shown in Fig. 2-20, Cruise Performance.

For example, let us assume that you have selected 6,000 feet ASL as the altitude for your cross-country flight (for illustration purposes, disregard the VFR Cruising Altitude requirement). Let us also assume that you are going to attempt to get the best range possible under the circumstances (consider the wind to be calm and the temperature to be standard).

By referring to the Range Profile chart (Fig. 2-19), you find that 45 percent power will provide the best range for 6,000 feet, slightly more than 420 NM. Now that you know how far you can go, you need to know your true airspeed in order to calculate your ground speed and ETA. To do this, consult the Cruise Performance chart (Fig. 2-20).

Find the 6,000 ft. Pressure Altitude figure on the left of the chart, and move across to the Standard Temperature column. Move down the % BHP column to the figure 45, read the first column to the right and find that your TAS will be 79 KT. As there is no wind, ground speed and true airspeed will be the same. Note that the column next to the Pressure Altitude provides the RPM required to achieve various power settings. In this case for 45 percent power you would adjust the throttle to 2000 RPM.

While there are occasions when you might want to achieve the best range possible for a given set of circumstances, the truth of the matter is that in the majority of cases you will probably trade the economy of flight for range (and the lower speeds) just to get to your destination more quickly.

The desire to get there faster is usually a matter of personal preference. If this should be your choice, the Cruise and Range charts would be used in a different order. After selecting a power setting and corresponding TAS from the Cruise Performance chart, you would then refer to the Range chart to determine how far you could safely go before stopping for fuel. A reminder once again, these charts do not make allowances for wind and in some cases, non-standard atmosphere.

Flight For Endurance

There is a maximum length of time an aircraft may remain airborne for a given power setting depending on the amount of fuel carried. This time or endurance will obviously be less when the aircraft is operated at high power settings and greater at low power settings.

However, when an aircraft is being operated in a manner that will enable it to remain in the air as long as possible for the amount of fuel carried on board, it is said to be flying for maximum endurance. While most pilots will seldom encounter situations necessitating flight for maximum endurance, there are circumstances that do require the use of this skill.

The pilot who requests Special VFR into a busy Class C aerodrome may be required to hold clear of the control zone due to IFR and possibly other SVFR arrivals and departures. Setting the aircraft up for maximum endurance flight might be advisable in this situation, especially if the possibility exists for a lengthy wait for a clearance into the zone.

Even in situations where a delay won't be so long as to threaten your fuel reserves, it makes sense to slow down and save fuel.

Perhaps the most important decision the pilot has to make is whether to set the aircraft up for endurance flight and wait for the situation to improve, or take

CRUISE PERFORMANCE

CONDITIONS:
1,670 lbs.
Recommended Lean Mixture

PRESSURE ALTITUDE FT	RPM	20° BELOW STANDARD TEMP			STANDARD TEMPERATURE			20° ABOVE STANDARD TEMP		
		% BHP	KTAS	GPH	% BHP	KTAS	GPH	% BHP	KTAS	GPH
2,000	2400	---	---	---	75	101	6.1	70	101	5.7
	2300	71	97	5.7	66	96	5.4	63	95	5.1
	2200	62	92	5.1	59	91	4.8	56	90	4.6
	2100	55	87	4.5	53	86	4.3	51	85	4.2
	2000	49	81	4.1	47	80	3.9	46	79	3.8
4,000	2450	---	---	---	75	103	6.1	70	102	5.7
	2400	76	102	6.1	71	101	5.7	67	100	5.4
	2300	67	96	5.4	63	95	5.1	60	95	4.9
	2200	60	91	4.8	56	90	4.6	54	89	4.4
	2100	53	86	4.4	51	85	4.2	49	84	4.0
	2000	48	81	3.9	46	80	3.8	45	78	3.7
6,000	2500	---	---	---	75	105	6.1	71	104	5.7
	2400	72	101	5.8	67	100	5.4	64	99	5.2
	2300	64	96	5.2	60	95	4.9	57	94	4.7
	2200	57	90	4.6	54	89	4.4	52	88	4.3
	2100	51	85	4.2	49	84	4.0	48	83	3.9
	2000	46	80	3.8	45	79	3.7	44	77	3.6
8,000	2550	---	---	---	75	107	6.1	71	106	5.7
	2500	76	105	6.2	71	104	5.8	67	103	5.4
	2400	68	100	5.5	64	99	5.2	61	98	4.9
	2300	61	95	5.0	58	94	4.7	55	93	4.5
	2200	55	90	4.5	52	89	4.3	51	87	4.2
	2100	49	84	4.1	48	83	3.9	46	82	3.8
10,000	2500	72	105	5.8	68	103	5.5	64	103	5.2
	2400	65	99	5.3	61	98	5.0	58	97	4.8
	2300	58	94	4.7	56	93	4.5	53	92	4.4
	2200	53	89	4.3	51	88	4.2	49	86	4.0
	2100	48	83	4.0	46	82	3.9	45	81	3.8
12,000	2450	65	101	5.3	62	100	5.0	59	99	4.8
	2400	62	99	5.0	59	97	4.8	56	96	4.6
	2300	56	93	4.6	54	92	4.4	52	91	4.3
	2200	51	88	4.2	49	87	4.1	48	85	4.0
	2100	47	82	3.9	45	81	3.8	44	79	3.7

Figure 2-20   Cruise Performance

other action. Proceeding to an alternate may be the most prudent course of action in many cases.

Altitude

For reciprocating engines, maximum endurance is achieved at sea level. That is not to say that you must fly at sea level each time you set the aircraft up for endurance flight. However, the best endurance performance will be achieved at the lowest practicable altitude commensurate with safety, which includes considerations such as traffic, ATC instructions or clearances, and obstacles.

Turbulence

Turbulence will have a significant effect on the endurance performance of an aircraft. This is due to the upsetting effect turbulence has on the stability of the aircraft.

Turbulence changes the angle at which the relative airflow meets the wing. At higher angles of attack, more power is required to overcome the increased drag. Because the aircraft is already being operated at the minimum power setting to maintain level flight, there will not be sufficient power to both overcome the increased drag and maintain altitude.

Therefore, if the pilot is to maintain level flight, it will be necessary to constantly change the power setting resulting in a considerable increase in fuel consumption. In this case, the pilot would be better off using a slightly higher continuous power setting.

Flaps

Flaps are not generally used when the aircraft is being flown for endurance. While increasing the coefficient of lift, flaps will also increase the coefficient of drag resulting in more power being required to maintain level flight thus defeating the purpose.

Mixture

Once the minimum power setting for level flight has been found, maximum endurance may only be achieved with the mixture leaned properly. You may notice that as the mixture is leaned, in the case of an aircraft with a fixed pitch propeller, there will be an increase in RPM. If this should be the case, remember to readjust the RPM.

Determining Endurance

Endurance flight may be determined by reference to the information provided in the Aircraft Flight Manual, usually in the form of a chart, or by experimenting with various power settings.

Using the Chart

Most manufacturers do not provide a particular speed for maximum endurance. The Endurance Profile chart shown in Fig. 2-21, is typical of the endurance information provided by one manufacturer and simply gives the endurance that can be expected at various power settings. Read the conditions associated with the chart carefully.

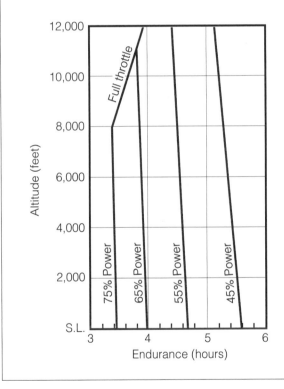

Endurance Profile
45 minutes reserve
24.5 gallons usable fuel

Conditions:
1670 lbs
Recommended lean mixture for cruise
Standard temperature

Notes:
1. This chart allows for the fuel used for engine start, taxi, takeoff and climb, and the distance during climb.
2. Reserve fuel is based on 45 minutes at 45% BHP and is 2.8 gallons.

Figure 2-21 Endurance Profile

Use of this chart is straightforward. Select an altitude from the left side of the chart and move horizontally to the right to the desired power setting. From the intersection of these two lines, move vertically down to the figures at the bottom of the chart representing Endurance (hours). The increments between these numbers represent .2 hours or 12 minutes. For example, at 4,000 feet and 65 percent power, endurance for this aircraft is just under 4 hours.

Experimental Method

Referring to the Endurance Profile chart shown in Fig. 2-21, note that the lowest power setting provided by the chart is 45 percent, giving an endurance of 5.6 hours at sea level. However, 45 percent power may not be the lowest power setting required to maintain level flight. In fact, most light aircraft are capable of maintaining level flight at much lower power settings.

Why then doesn't the manufacturer provide a specific power setting or airspeed for maximum endurance? As with range, it is impossible to do so due to the number of variable factors, such as altitude, temperature, and gross weight, that may affect the endurance performance of an aircraft on a given flight.

In the absence of any specific information in the Aircraft Flight Manual, it is left to the pilot to determine the power setting for maximum endurance for a given set of circumstances. You may find the approximate maximum endurance speed for your aeroplane experimentally by flying at various power settings while maintaining level flight.

For example, commencing at a power setting in the mid-cruise range for your aeroplane, decrease power in 100 RPM increments while retrimming for level flight and allowing the airspeed to stabilize at a constant value. You will eventually arrive at a point where further reductions of power will result in a stall if you attempt to maintain altitude, or in a loss of altitude if you attempt to prevent the stall. In fact, the only way you can prevent either of these events from occurring is to increase power so that you will now find yourself flying at an airspeed requiring a somewhat higher power setting to maintain level flight. You have now entered slow flight. (See Fig. 2-18, Power/Drag Curves.)

It is now a matter of adjusting power to the lowest setting that enables you to maintain level flight, and you will be flying for endurance. Remember to retrim the aircraft.

Fig. 2-22, shows an actual test of an aeroplane with a fixed pitch propeller and fixed gear with a power-off stall speed of 56 KT. Note that for this particular aircraft, the minimum power setting at which level flight is possible is 2075 RPM which in this case, corresponds to a speed of 65 KT indicated airspeed. In this manner, the power setting for maximum endurance for any aircraft with a reciprocating engine may be determined.

RPM Indicated Airspeed (KT)

RPM	Indicated Airspeed (KT)
2700	101
2600	96
2500	90
2400	86
2300	78
2200	74
2150	70
2075	**65**
2250	61
2300	57

Figure 2-22 Endurance Power and Resultant Airspeed

The power setting for endurance flight determined in this manner, will be adequate for straight-and-level flight. However, if turns are to be made, and this is usually the case when you are "holding," it will be necessary to increase power slightly to maintain altitude and compensate for the forces acting on the aircraft in the turn. Leaving the power constant at this setting will result in less fuel being used than if the power is changed each time you turn.

Remember, the decision to fly for maximum endurance while waiting for a particular situation to improve must be an informed one based on the knowledge that all of the factors that influence good decision making have been considered and that reasonable options exist to provide an out if things don't go as planned. If in doubt, don't hesitate to proceed to an alternate.

Slow Flight

Slow flight, for the purposes of this manual, may be defined as, "that range of airspeeds between the maximum endurance speed for a particular aircraft and the point just above its stalling speed for the existing flight conditions." Training in slow flight has four main purposes:

1. To learn to recognize the symptoms when approaching the slow flight speed range to avoid inadvertent entry into this speed range.
2. To maintain safe flight control, in all configurations, within the slow flight speed range. This will aid in development of co-ordination and instill confidence in the handling of the aircraft.
3. To acquaint the student with the possible consequences of failing to take prompt corrective action, particularly when flying at airspeeds close to minimum control speed.
4. To learn to recover to normal airspeeds promptly with a minimum loss of altitude.

There are several conditions where an aircraft may encounter slow flight. Some of these conditions are: take-offs, landings, recovering from a misjudged landing, and an approach to a stall. Of prime importance is your awareness of the effect of controls in slow flight.

During slow flight, control and management of the aircraft require the full attention of the pilot. Furthermore, operation in slow flight is not necessarily economical: fuel consumption is higher and engine damage can result from overheating during prolonged flight at these airspeeds, particularly while attempting to climb. An aircraft should not be operated in slow flight while waiting for weather to clear, while inspecting a potential landing area, or while searching for ground fixes when lost.

When operating within the slow flight speed range, you must know and understand the characteristics associated with the performance and control of the aircraft, especially the control of altitude, airspeed, and yaw.

To enter slow flight, first establish the aircraft in straight-and-level flight for maximum endurance. Once established, raise the nose beyond the normal nose-up limits for maximum endurance. The airspeed will decrease due to increased drag, and a loss of height will become apparent. To offset the loss of height and to maintain altitude, an increase in power will be required. The aircraft is now in slow flight (Fig. 2-23).

With the aircraft established in slow flight, a further decrease in airspeed without a change in power setting will result in a loss of height. Therefore, you must increase power to maintain a constant altitude. Conversely, an increase in airspeed without a change in power setting will result in a decrease in induced drag. You must then reduce power to maintain a constant altitude at the higher airspeed. There may be a slight loss of altitude during the transition period.

Slow flight at minimum airspeeds should be practised in straight-and-level flight, climbing and descending, level turns, and climbing and descending turns. Should it be necessary to manoeuvre in slow flight in gusty wind conditions, the airspeed must be adjusted upward to allow for the gust factor.

Dual flight instruction should include exposure to slow flight in a climb to simulate the conditions that may be experienced during an overshoot at a high density altitude, or when affected by obstacles, to demonstrate the need for prompt and proper transition from slow flight.

When you are established in a constant rate of descent in slow flight, a reduction in airspeed will result in an increased rate of descent, thereby necessitating an increase in power to maintain the former rate of descent.

It should be remembered that slow flight should be of short duration because of the insufficient airflow for engine cooling.

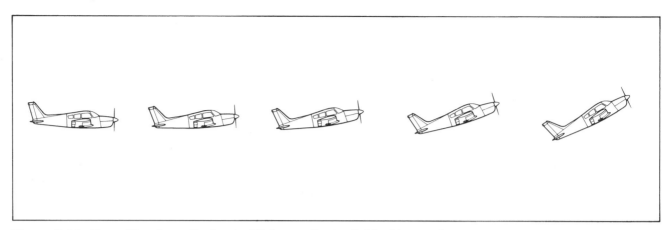

Figure 2-23   Transition from Cruise to Minimum Controllable Airspeed

Flaps

Straight-and-level flight, climbs, descents, and turns in slow flight should be practised at various flap settings. Your instructor will show you how certain flap settings may afford better forward visibility due to a lower nose attitude. You will also see how flap extension will decrease the stall speed; thus, allowing you to fly safely at lower speeds. Flaps will often be extended when it is necessary to transition from slow flight to level flight or a climb: examples are recovering from a bad landing and an overshoot. Therefore, you should become proficient in controlling the aircraft in slow flight at different flap settings.

Control Responses

The control responses in slow flight will be different from those experienced when flying in the normal speed range. Lack of aileron response will be the most noticeable, and the lower the airspeed the less effective the ailerons will be. In fact aileron drag on most aeroplanes will become quite prominent in slow flight and produce a noticeable yaw opposite to the direction of intended turn. When entering a turn during slow flight, you will have to compensate for aileron drag by use of rudder in the direction of the desired turn.

Elevators and rudder will be influenced by the propeller slipstream, and these controls will continue to be effective. Any alteration in power or speed will affect all the controls and changes in response may be noticed.

It is very important to note that during slow flight the engine and propeller produce slipstream and asymmetric thrust, which tend to yaw the aeroplane to the left. Remember that this yaw must be controlled by rudder. Depending upon the power and speed combination, a firm right rudder pressure may be required to maintain direction. As well, because of the high power low airspeed combination, torque will tend to roll the aeroplane to the left. This rolling is not as noticeable as the yaw, but nevertheless must be compensated for with aileron.

Encountering A Stall

As speed is further reduced, more power is required to maintain the desired height (or rate of descent), but when the nose pitches down with full power applied, the airspeed has gone below the minimum controllable airspeed. At this point the aircraft has entered a stalled flight condition. Near the stalling speed, factors such as air density, aircraft weight, and the drag from flaps and landing gear (when applicable) may produce a condition in which it becomes impossible to maintain height. This may happen in attempting a turn after take-off with a high all-up weight and a critical density altitude.

The slow flight speed range does not automatically imply serious control difficulties or hazardous conditions. However, it does amplify any errors of basic flying technique. Hence, proper technique and precise control of the aircraft are essential.

Stalls

A stall is a loss of lift and increase in drag that occurs when an aircraft is flown at an angle of attack greater than the angle for maximum lift. Stall training will allow you to recognize the symptoms of an approaching stall early enough to take action to prevent a stall from happening. You will also learn how to recover positively and smoothly with a minimum loss of altitude should a stall occur.

Why Does a Wing Stall?

The lift generated by a wing is dependent upon a smooth accelerated airflow over the wing. At moderate angles of attack the airflow near the trailing edge of the wing becomes mildly turbulent. As the angle of attack increases, the turbulent air progresses forward toward the leading edge of the wing until the stalling angle is reached. At that point, the downwash and the pressure differential are greatly reduced, and a loss of lift results. Due to the loss of lift and increase in drag, the remaining lift is insufficient to support the aeroplane, and the wing stalls. The four parts of Fig. 2-24 show the airflow over the aerofoil at various stages leading to the stall.

It is basic in recognizing stalls to remember that, unlike angle of incidence, angle of attack is a relative factor. Therefore you cannot rely upon aircraft attitude entirely to indicate the possibility of a stall. Angle of attack may be simply defined as the angle between the mean chord of an aerofoil and its direction of motion relative to the airflow (relative airflow). In this

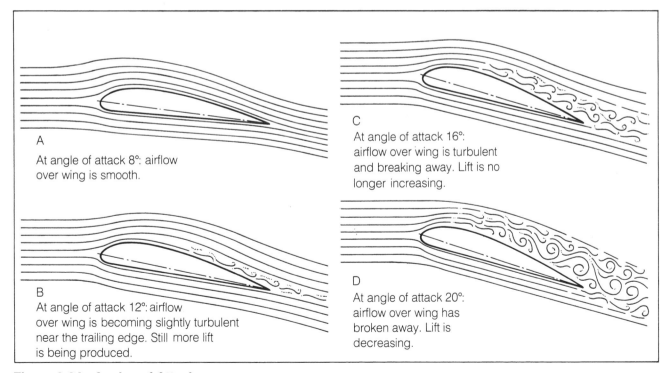

A
At angle of attack 8°: airflow over wing is smooth.

B
At angle of attack 12°: airflow over wing is becoming slightly turbulent near the trailing edge. Still more lift is being produced.

C
At angle of attack 16°: airflow over wing is turbulent and breaking away. Lift is no longer increasing.

D
At angle of attack 20°: airflow over wing has broken away. Lift is decreasing.

Figure 2-24   Angles of Attack

manual, the term "relative airflow," is used to describe the direction of the airflow with respect to an aerofoil in flight. An aircraft may be stalled in practically any attitude and at practically any airspeed.

Stalling Speeds

Regardless of airspeed, an aircraft always stalls when the wings reach the same angle of attack. Remember, angle of attack and aircraft attitude are not consistently related. Although stalling speeds may be given for a specific type of aircraft, the stalling speed for each aircraft may vary with the following factors:

Weight.   Since weight opposes lift, a lightly loaded, properly balanced aircraft will have a lower stalling speed than a similar aircraft operating at its maximum permissible weight.

Balance.   The position of the Centre of Gravity will also affect the stalling speed of an aircraft. A forward Centre of Gravity location will cause the stalling angle of attack to be reached at a higher airspeed while a rearward Centre of Gravity will cause the stalling angle of attack to be reached at a lower airspeed.

An improperly loaded aircraft may display undesirable stalling characteristics. This is particularly true of an aircraft loaded beyond the aft Centre of Gravity limits.

Power.   Because of the additional upward thrust and other lift contributing factors of a power-on stall, the stalling speed will be lower than with power off.

Flaps.   When flaps are extended the camber of the wing is effectively increased. This deflects more of the airflow downward for a given airspeed, thereby increasing lift. This factor allows the aircraft to be flown at a lower speed before the stall occurs.

Pitch.   When an aircraft is pitched upward abruptly, the load factor is increased correspondingly and a higher stalling speed is introduced for the duration of change in pitch attitude (see Chapter 1, "Load Factor").

Angle of Bank.   The greater the bank angle, in co-ordinated flight, the higher the stalling speed.

Aircraft Condition.   A clean, well-maintained, properly rigged aircraft will invariably have better stalling characteristics and lower stalling speeds than a similar aircraft in poor general condition.

Retractable Landing Gear.   Extending the landing gear increases drag. The effect on stalling speed varies from aircraft to aircraft, but generally in the classic wings level nose-up attitude a slightly lower stalling speed will be noted, especially in the power-on configuration.

With altitude, the density of the air in which an aircraft is flying decreases. Although the true airspeed at which the aeroplane stalls is higher at altitude, the airspeed indicator, which itself functions by the effect of the air density, will record the same speed when the aircraft stalls at altitude as it did at or near ground level. Therefore, indicated stalling speeds will remain the same at all altitudes.

Symptoms of an Approaching Stall

As most aircraft do not have an angle of attack indicator, the airspeed must be used as a guide to identifying the approach to a stall. Other symptoms of an approaching stall are:

1. A decrease in the effectiveness of the controls, especially elevator and aileron control: the "live" resistance to pressures on the controls becomes progressively less and less as speed decreases.
2. Audible or visual stall warning devices fitted in most aircraft are activated prior to the stall.
3. Buffeting (the beating effect of turbulent airflow on the aircraft's structure, which can be heard and/or felt) varies in intensity with different types of aircraft.
4. Loss of height, despite rearward movement of the elevator control.

Always remember that should the approach to stall symptoms begin when a stall is not intended, movement of the controls should be smooth and prompt. Most aerofoils stall at about 17 degrees angle of attack; therefore, the stall symptoms occur at 15 to 16 degrees. The alert pilot will be able to recognize the stall since it normally occurs gradually. Because of the wash-out of the wings, the stall begins at the wing roots, and, as the angle of attack is increased, moves progressively toward the wing tips. When the first symptoms of a stall occur, move the elevator control forward smoothly and promptly to reduce the angle of attack and to return the aircraft to stabilized flight. If additional power is available you may effectively reduce the angle of attack by applying appropriate power, without any change in the aircraft's pitch attitude. However, under normal conditions you should eliminate stall symptoms by adding available power and lowering the nose to reduce the angle of attack. Approaching close to the stall but not fully stalling is referred to as an imminent stall (Fig. 2-25).

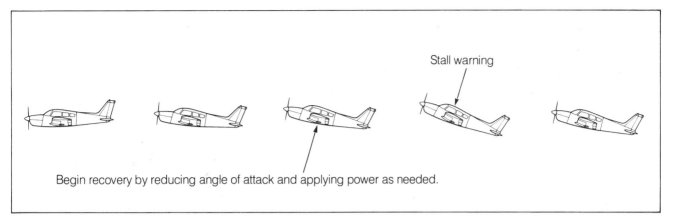

Stall warning

Begin recovery by reducing angle of attack and applying power as needed.

Figure 2-25   Recovery from an Imminent Stall

Power-Off Stall Entry and Recovery

Practise stalls only over an unpopulated area and at an operationally safe altitude. Practice stall recoveries should be completed at or above the height recommended by the manufacturer, or no less than 2,000 feet above ground, whichever is the greater. During training your instructor will emphasize repeatedly that the objective is not how to stall an aircraft but how to recognize the onset of a stall condition and take prompt corrective action.

Intentional stalls must be preceded by:

1. The cockpit check. Check for such things as carburettor ice, seat-belts secure, windows shut, loose articles secured, etc.
2. The look around. Do a very careful look around in all directions, especially below.

Power-off stalls are generally entered from straight-and-level flight. Apply carburettor heat if required, close the throttle smoothly, control yaw with rudder, and hold the aircraft in level flight by continued back pressure on the control column. The airspeed will gradually decrease through endurance speed and enter the slow flight range. As the airspeed decreases, control response diminishes and greater displacement of the controls will be required to produce a desired result. You will also notice a change in tone and intensity of the slipstream noise. Maintain the back pressure to hold the nose in the attitude used in a normal climb until the stall occurs. Keep straight with rudder control.

When practicing stalls, avoid raising the nose of the aircraft too far above the horizon to produce the stall (Fig. 2-26). This will ensure you experience more realistic symptoms of an accidental stall.

A simplified explanation of the procedure for recovering from a stall is:

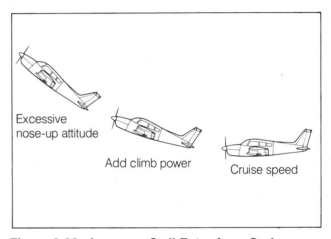

Excessive nose-up attitude

Add climb power

Cruise speed

Figure 2-26   Improper Stall Entry from Cruise

1. Immediately reduce the angle of attack.
2. Regain a correct flight attitude with co-ordinated use of flight controls. Not every situation demands the application of power, but its use must become an integral part of the recovery procedure (Fig. 2-27).

First, at the indications of a stall lower the nose positively and immediately. The amount of forward control movement required varies from aircraft to aircraft — in most cases only moderate pressure to about the neutral position is needed. The objective is to reduce the angle of attack sufficiently to smooth the airflow over the wing. Keep straight with rudder. Use ailerons if permitted by the Aircraft Flight Manual.

Second, apply full power promptly and smoothly. Rough or abrupt throttle operation may result in delayed engine response or a complete loss of power at a most crucial moment, especially if it has cooled off during a glide. At altitude a cruise power setting is normally sufficient, but at low altitudes application of full power is necessary to minimize loss of altitude. Due to the low airspeed there is very little risk of

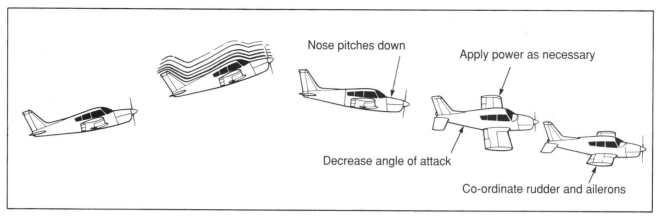

Figure 2-27   Full Stall Recovery

overspeeding the engine. It's better to use too much power at this point than too little. Resume the normal power setting as soon as recovery is accomplished, and you have regained cruise attitude.

Straight-and-level flight is regained by co-ordinated use of controls. Recovery from the stall should result in regaining normal flight with the least loss of height. In the average light aircraft, with the use of power, lowering the nose to the cruise attitude, or slightly below, will accomplish this end.

A stall can be aggravated by yaw. As the aeroplane yaws, a difference in lift between the two wings develops causing one wing to drop. This automatic rolling tendency is called *autorotation* and is described in detail in Chapter 13.

Should the aircraft yaw and a wing drop, carry out the full stall recovery procedure: unstall the aircraft, keep straight by controlling yaw, and level the wings with co-ordinated use of flight controls (Fig. 2-28).

Power-On Stall Entry and Recovery

The principles that apply to power-off stalls also apply to stalls entered with power, although there are some differences in the manoeuvres. The pitching of the aircraft from a full stall with power on is much more steep and rapid. The aircraft is also more difficult to control during recovery, since in many cases there is a tendency for one wing to drop at the same time as the nose pitches down.

When you enter a stall with power on, the elevators and rudder retain their effectiveness longer due to the propeller slipstream. However, the ailerons are less effective than in a power-off stall. This is partially due to power causing the stalling speed to be slightly lower, which decreases aileron effectiveness.

To enter a stall with power on, raise the nose smoothly to a nose-up attitude and hold it there with continued aft movement of the control column as necessary until the stall occurs. Because of the additional thrust the nose must be raised higher to accomplish a stall. During a power on entry, yaw due to slipstream and asymmetric thrust will make directional control more difficult.

Recovery from a power-on stall with any given power setting is made in the same manner as recovery from any other stall. However, as the stall could be more abrupt and the loss of control more complete, full application of any remaining power is of great importance.

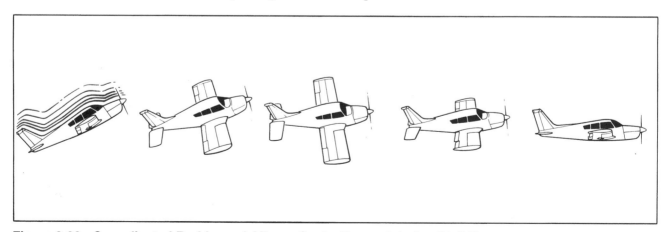

Figure 2-28   Co-ordinated Rudder and Aileron Control is used during Stall Recovery

Acceleration Stalls

At the same gross weight, configuration, and power setting, an aircraft will consistently stall at the same indicated airspeed, provided no additional load factor is incurred by a manoeuvre or abrupt use of the controls. The aircraft will, however, stall at a higher airspeed when manoeuvring loads are imposed by sudden turns, pull-ups, or abrupt changes in its flight path. Stalls entered from such flight situations are called *acceleration stalls*.

Following an intentional stall, if the nose has been allowed to pitch down at a large angle the airspeed will increase rapidly. Should the elevator control be brought back too rapidly to recover from the dive the aircraft will enter a "secondary stall" at a much greater speed than that experienced in a normal stall. Regardless of airspeed trend, any movement of the controls which increases the "G" factor produces an acceleration.

In a turn, the aircraft is accelerated toward the centre of the turn; the steeper the turn, the greater the acceleration, hence the greater the load factor and the higher the stalling speed. An aircraft that might normally stall at 50 KT may stall at a speed well in excess of 100 KT in a steep turn when the angle of bank exceeds 60 degrees. Therefore, acceleration forces generated by turns or abrupt changes in upward pitch, regardless of airspeed, will always increase the stalling speed. This holds true in climbing turns, level turns, gliding turns, steep turns — in fact any turn, regardless of power. A manoeuvre which requires special attention to airspeed and angle of bank is the power-off (gliding) turn, especially when conducted close to the ground.

Turbulence can cause a significant increase in stalling speed. An upward gust causes an abrupt change in the relative airflow, which results in an equally abrupt increase in angle of attack. All stalls are caused by exceeding the critical angle of attack, and the base from which this angle is measured is the direction of motion of the relative airflow. This is why an airspeed slightly higher than normal is usually recommended when approaching to land in turbulent conditions.

That an aircraft may be stalled at any airspeed, does not imply that it is permissible to do so. The lower the airspeed when a stall occurs the less the possibility of structural damage, as the load factors are minimal. However, the higher the airspeed the higher the load factor, and correspondingly the greater the chance of inflicting structural damage. This is why Aircraft Flight Manuals include "manoeuvring speed" as an operating limitation. Manoeuvring speed is the maximum speed at which the application of full aerodynamic control will not overstress the aircraft.

Stalls During Turns

When an aircraft is stalled during a level or descending turn, the inside wing normally stalls first, and the aircraft will roll to the inside of the turn. In a level turn, the inside wing is travelling more slowly than the outside wing and obtains less lift, causing it to sink and increase its angle of attack. Under the proper conditions, this will produce a stall. During a descending turn, the path described by the aircraft is a downward spiral; therefore, the inside wing is meeting the relative airflow at a steeper angle of attack and is the one to stall first and drop lower.

However, during a climbing turn, the path described by the aircraft is an upward spiral; therefore, the outside wing is meeting the relative airflow at a steeper angle of attack than the lower wing. As a result, the higher wing will normally stall first and drop abruptly when the stalled condition occurs.

Departure Stalls

During take-off and the initial stage of departure, an aircraft enters into and passes through a critical condition of flight. After leaving the ground and accelerating to climbing airspeed, the aircraft passes through a period of low airspeed at low altitude. Any abrupt pull-up or reduction in engine power could cause the aircraft to stall. Should a mishap occur at this point and good airmanship has prevailed, the throttle can be closed and a landing safely made straight ahead with only small changes in direction to avoid obstructions. However, should an aircraft attitude become too nose high after rotation, a stall may occur from which a successful recovery cannot be made, or if the aircraft is in a near stalled condition, it will not climb sufficiently to clear obstacles in the flight path (Fig. 2-29). Therefore, establishing the correct nose-up attitude for a climb after take-off is imperative. As part of the departure procedure, take great care to establish the correct nose-up attitude when executing a climbing turn, especially if the turn must be carried out before a safe height is reached.

Another critical departure procedure is the "overshoot" resulting from a missed approach. More often than not, when a decision is reached to discontinue an approach, the airspeed is low and the flaps are extended. In addition, it may be necessary to turn very shortly after initiating the missed approach (for traffic pattern purposes). To avoid conditions that may lead to a stall or near stall, pay particular attention to the following:

1. Apply full power. Remember, this is a form of

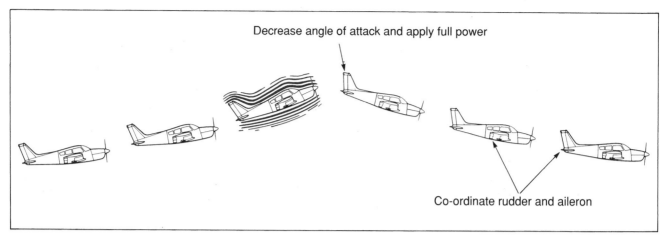

Decrease angle of attack and apply full power

Co-ordinate rudder and aileron

Figure 2-29 Departure Stall

take-off under adverse conditions so nothing less than full power is adequate.

2. Application of power plus the nose-up trim used during the approach will tend to force the aircraft into a nose-high attitude. Anticipate this and compensate by holding the correct pitch attitude until the trim can be readjusted.

3. Very few aircraft are able to sustain a climb with flaps fully extended. Retract the flaps smoothly in accordance with the instructions in the Aircraft Flight Manual. Should the Aircraft Flight Manual not indicate how to raise the flaps, it is recommended that they be raised in stages. When the flaps are fully retracted immediately, a sudden loss of height can occur. Attempts to arrest this descent by raising the nose suddenly may induce a stall.

Spinning

The *spin* has no practical application in normal flight. You are trained in spins to learn recognition, avoidance, and recovery.

It is imperative that only aircraft certified for intentional spinning be used for any form of spin training. The type certificate, Aircraft Flight Manual, or cockpit placards must be consulted to determine under what conditions, if any, spin practice may be undertaken in a particular aircraft. The pilot of an aircraft placarded against intentional spins should assume that it might become uncontrollable in a spin. Entry and recovery techniques recommended in this text apply to the average light training aircraft. Should the Aircraft Flight Manual dictate different techniques, they must be followed.

Autorotation

An automatic rolling tendency, or autorotation, will develop following a stall that has been aggravated by yaw. If allowed to continue, autorotation will develop into a spin.

If for some reason one wing of an aircraft produces more lift than the other, the aircraft will roll and the downgoing wing will meet the relative airflow at a greater angle of attack. At ordinary angles of attack, this increase in angle will cause an increase in lift, which tends to restore the aircraft to its previous attitude. When a downgoing wing experiences an increased angle of attack at or near the stalling angle,

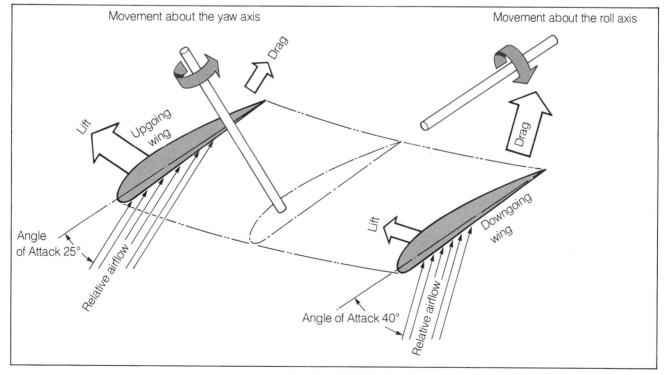

Figure 2-30　Angles of Attack and Forces Acting in a Spin

it loses lift, becomes more stalled, and automatically continues to drop. The upgoing wing, because of its relative upward movement, meets the airflow at a reduced angle of attack, becomes less stalled, and produces more lift, which accentuates the roll. Also, drag on the downgoing wing increases sharply, adding to the existing yaw force, which effectively increases the angle of attack of the downgoing wing, stalling it further (Fig. 2-30). The nose drops owing to the loss of lift, and autorotation or spinning sets in.

The spinning motion is complicated and involves simultaneous rolling, yawing, and pitching. The aircraft follows a helical or corkscrew downward path, rotating about a vertical axis. Pitch attitudes may vary from flat to steep, while forward and vertical speeds are both comparatively low. Forces are somewhat above normal but are relatively steady.

The spin consists of the three stages illustrated in Fig. 2-31:

1. The *incipient* stage.
2. The *fully developed* stage.
3. The *recovery*.

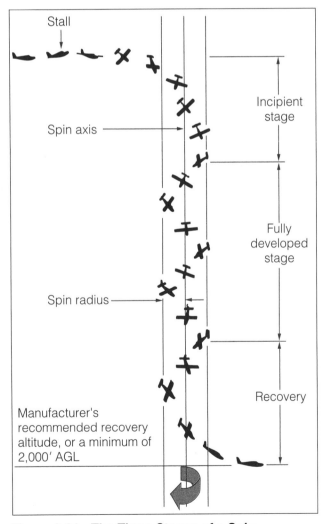

Figure 2-31   The Three Stages of a Spin

The Incipient Stage

The incipient stage occurs from the time the aircraft stalls and rotation starts until the spin axis becomes vertical or nearly vertical. During this time, the flight path changes from horizontal to vertical, and spin rotation increases from zero to the fully developed rate. The incipient stage usually occurs rapidly in light aircraft, some 4 to 6 seconds, and consists of approximately the first two turns. Model and actual tests show that the typical incipient stage motion starts during the stall with a wing drop. As the nose drops the yawing motion begins to increase. About the half-turn point, the aircraft is pointed almost straight down but the angle of attack is usually in excess of that of the stall, because of the inclined flight path (Fig. 2-31). Near completion of the first turn, the nose may come back up, and the angle of attack continues to increase. As the aircraft continues to rotate into the second turn, the flight path becomes more nearly vertical, and the three spinning motions become more repeatable and approach those of the fully developed stage.

The Fully Developed Stage

In the fully developed stage, the attitude, angles, and motions of the aircraft are somewhat repetitious and stabilized from turn to turn with a nearly vertical descent. The spin is maintained by a balance between the aerodynamic and inertia forces and moments (Fig. 2-32).

Entry

A spin, whether deliberate or inadvertent, may be entered in many ways. It is not necessary for an aircraft to have a relatively high pitch attitude for it to stall and spin. The angle of attack is the key factor, not the attitude. It is possible to enter a spin with the aircraft in a descending, level, or climbing attitude. A spin can also be entered from an accelerated stall. Many types of aircraft require special techniques to get the spin properly started. Strangely enough, these same aircraft have been known to spin accidentally, due to mishandling in routine turns or in slow flight.

The primary requirement is that the aircraft be fully stalled, otherwise it might not spin, and the result would likely be a skidding spiral of increasing airspeed. If this occurs, immediately recover from the spiral dive and start over.

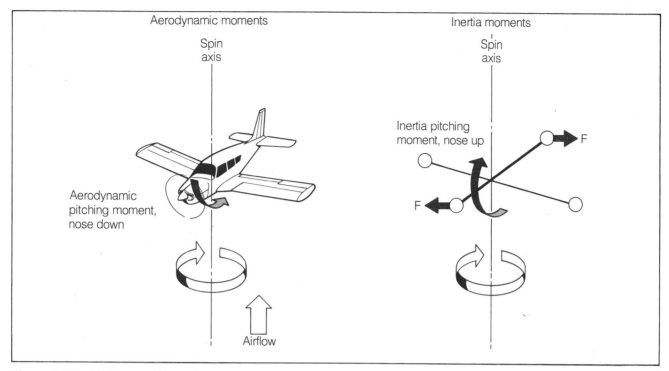

Figure 2-32   Balance of Aerodynamic and Inertia Pitching Moments in a Spin

One method of inducing a spin is outlined below.

1. Complete safety precautions — cockpit checks, minimum altitude, suitable area, look-out, etc. All practice spin recoveries should be completed no less than 2,000 feet above ground, or at a height recommended by the manufacturer, whichever is the greater.
2. Reduce power to a minimum and stall the aircraft by gradually applying full aft control column while maintaining a near normal climb attitude. At or slightly before the stall, apply full rudder in the direction of the desired spin.
3. Allow autorotation to occur by maintaining full rudder and holding the control column fully back, as at this point there may be an instinctive tendency to release pressure.
4. Allow the spin to progress through the desired number of turns, but never through more than six. Approved aircraft are not tested beyond these limits. Normally, two complete turns of a developed spin should be sufficient.

Recovery

The aim in recovery is to upset the balance between the aerodynamic and inertia moments. Because aircraft spin characteristics differ, recovery techniques specified in the Aircraft Flight Manual must be followed. The procedures outlined below are suitable for most small aircraft and may be used in the absence of manufacturer's data.

1. Power to idle, neutralize ailerons.
2. Apply and hold full rudder opposite to the direction of rotation.
3. Just after the rudder reaches the stop, move the control column positively forward far enough to break the stall. Full down elevator might be required.
4. Hold these control inputs until rotation stops.
5. As the rotation stops, neutralize rudder, level the wings, and recover smoothly from the resulting dive.

Factors Affecting Recovery

The most important difference between the fully developed stage and the incipient stage is an increase in recovery time, for some aircraft, and to a lesser extent the amount of control input needed. From the fully developed stage it is not unusual for a full turn or more to occur after the application of recovery controls before rotation stops. Therefore, it is very important to apply the recovery controls in the proper sequence and hold them until rotation stops. Premature relaxation may extend the recovery time.

Some of the factors likely to affect spin behaviour and recovery characteristics are: aircraft loading (distribution, Centre of Gravity, and weight); altitude; power; flaps; and rigging.

Distribution of the weight in the aircraft can have a significant effect on spin behaviour. The addition of weight at any distance from the Centre of Gravity of the aircraft will increase its moment of inertia about two axes (Fig. 2-32). This increased inertia, independent of the Centre of Gravity location or weight, will tend to promote a less steep spin attitude and more sluggish recoveries. Forward location of the Centre of Gravity will usually make it more difficult to obtain a pure spin, due to the reduced elevator effectiveness. The farther back the Centre of Gravity is, and the more masses distributed along the length of the fuselage, the flatter and faster the spin tends to become. Changes in gross weight as well as in its distribution can have an effect on spin behaviour, since increases in gross weight will increase inertia. Higher weights may extend recoveries slightly.

High altitudes will tend to lengthen recoveries since the less dense air provides less "bite" for the controls to oppose the spin. However, this does not suggest you should use low altitudes for spin practice.

The effect of the use of ailerons, either with or against the rotation, apparently follows no set rule for all aircraft. As application of ailerons might increase the rotation rate and delay recovery, there must be no tendency to use ailerons, particularly in a cross-control manner.

If a spin occurs with flaps extended, retract them, as extended flap might:

1. Prolong the spin, because it induces a flatter spin attitude and lower spin rate.

2. Reduce the effectiveness of the rudder, due to deflected air flow.
3. Incur damage from high speed or high loading, or both, in recovery from the dive.

With power on, the attitude of the aircraft might be less nose down, and the propeller will tend to add some gyroscopic inputs, which will be reversed between left and right spins. The effect of leaving power on during a spin is to lengthen recoveries on some aircraft. Additionally, a power-on recovery will likely result in increased airspeed and height loss during the dive recovery.

If disorientation prevents determining the direction of rotation, refer to the turn needle or turn co-ordinator to establish the direction of rotation. For example, if the turn needle or turn co-ordinator indicates a turn to the left, the aircraft is spinning to the left. Do not refer to the ball indicator because the ball does not remain in a constant position, due to transient yaw.

Secondary Spin

A secondary spin may result from mishandling the controls following recovery from the initial spin. An abrupt or premature pull-up from the dive recovery could cause a secondary stall (Fig. 2-33). If yaw is present — for example, from inadvertent retention of anti-spin rudder — the aircraft might enter a secondary spin.

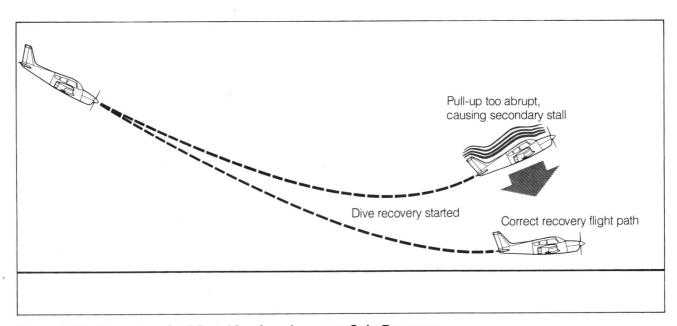

Figure 2-33 Secondary Stall Resulting from Improper Spin Recovery

Spiral

A spiral may be informally described as "a steep descending turn in which airspeed, rate of descent, and wing loading increase rapidly." The spiral is not usually considered a normal or useful manoeuvre, and in its accidental form it can become very hazardous.

It can be readily seen why this manoeuvre is considered hazardous, especially if it occurs at a low altitude. A high-speed stall could result from incorrect use of elevator in an attempt to check the rapid rate of descent. If airspeed is permitted to increase beyond normal limits, the aircraft can be structurally damaged during the spiral or during the pull-up from the dive if loading becomes excessive.

In a way, a spiral resembles a spin. Therefore, when executing practice spins it is possible to become temporarily disoriented, so that what appears to be a spin is actually a spiral. Under these conditions, always remember that the main difference between the two manoeuvres is airspeed. In a spin the airspeed is constant and low — at or about the stalling speed; in a spiral the airspeed will be well above stalling speed and increasing rapidly. A spiral may result from attempting to force an aircraft into a spin too soon before a stall occurs or from relaxing the elevator controls once a spin has started.

A spiral may also result from mismanagement of controls during manoeuvres in which additional engine power is used, such as steep turns. In these circumstances be sure that the throttle is closed, to bring the rapidly increasing airspeed under more effective control and to reduce the load factor build up when recovering from the dive.

Once recognized, it is not difficult to recover from a spiral dive. The following action must be taken promptly and in this order:

1. Close the throttle.
2. Roll the wings level. (Avoid rolling and pulling up at the same time.)
3. Ease out of the dive.
4. Apply power only after the airspeed has decreased to within the normal range.

The spiral is a manoeuvre that should not be practised solo, but for recognition and recovery action purposes, the flight instructor will demonstrate it in various ways, such as from an incorrectly entered spin or a poorly executed steep turn. During this demonstration, the sudden increase in airspeed and wing loading will be very evident. You will also see how any attempt to pull out of the dive without first levelling the wings further tightens and aggravates the manoeuvre.

Resist the temptation to roll the wings and pull out of the dive simultaneously, because there will be a much greater load imposed on the aircraft.

In carrying out intentional spirals the following points are stressed so that you can take adequate precautions:

1. Considerable height may be lost. Recovery should be completed at a height recommended by the manufacturer, or no less than 2,000 feet above ground, whichever is the greater.
2. Airspeed increases rapidly; take care not to exceed the speed limitations of the aircraft.
3. An attempt to recover from the ensuing dive too abruptly could result in an excessive load factor, with the danger of a pilot black-out, structural damage, or a high-speed stall.

Slipping

Slipping is a manoeuvre in which the aircraft is placed in a banked attitude but its tendency to turn is either reduced or prevented by the use of rudder.

Slipping is used for two purposes. One purpose is to increase rate of descent without increasing airspeed. For aircraft without flaps, this technique, known as a *forward slip*, is essential in controlling the angle of the approach. Even for aircraft equipped with flaps, slipping can still be used to correct the approach angle, provided that the Aircraft Flight Manual allows slipping while flaps are extended. Moreover, flaps have been known to fail. Another purpose of a slip is to counteract the effect of drift when landing in a cross-wind. This is called a *side-slip*.

The forward slip is one in which the longitudinal axis is at an angle to the desired descent path. If there is any cross-wind, the slip is more effective if made into the wind. As the aim is to increase the rate of descent, the slip should be done with the engine idling.

To enter a slip, use ailerons to lower the wing on the side toward which the slip is to be made. At the same time, use rudder to move the nose in the opposite direction. Bank applied one way is balanced by rudder applied the opposite way. The result is a constant direction of flight during the slip. Use elevators to maintain the desired airspeed. Note, however, that because of the location of the pitot tube and the static vents, slipping can cause airspeed errors. Pilots must learn to recognize a properly performed slip by the aircraft attitude and the feel of the flight controls.

Anticipate control pressures when maintaining a slip. If full rudder is used, considerable aileron pressure may be needed to maintain the bank. There will also be a tendency for the nose to pitch up as a result of the banked attitude and rudder input. This must be counteracted by the use of elevators to maintain the proper pitch attitude.

Recovery from a slip is achieved by simultaneously releasing rudder pressure, levelling the wings, and adjusting the pitch attitude to resume normal descent and airspeed.

A side-slip is a slip in which the aircraft's longitudinal axis remains parallel to the original flight path. This is essential in a cross-wind landing when the aircraft is slipped into the wind the right amount to counter the effect of drift. The desired flight path in this case is the centre of the runway. The aircraft must

Figure 2-34 Forward Slip

remain positioned over the centre of the runway while the longitudinal axis is held in alignment with it. This is a highly developed skill as control inputs must change as airspeed reduces and, if the wind is gusting, as the wind changes in speed and direction.

A *slipping turn* achieves the same aim as a forward slip — increased rate of descent without increasing airspeed — but does this in a turn. The turn is slowed, but not prevented, by the use of opposite rudder. The slipping turn can be useful during the turn to final approach, especially in the case of a forced landing in which excess altitude must be lost.

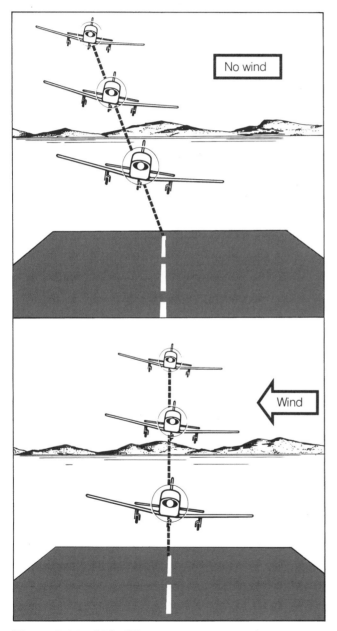

Figure 2-35   Side-Slip

Take-off

Taking off, as defined in the *Canadian Aviation Regulations* in relation to an aircraft, means "the act of leaving a supporting surface and includes the immediately preceding and following acts." In a normally executed take-off the aircraft becomes airborne smoothly and efficiently, with the minimum take-off run consistent with positive control and good climb performance. It is a requirement of the *Canadian Aviation Regulations* that take-offs at aerodromes be executed into wind, insofar as practicable, unless otherwise authorized by an appropriate Air Traffic Control unit. It makes good sense to take off into the wind since it:

1. Permits a shorter run and a lower ground speed at the moment of take-off.
2. Eliminates drift, so that there is no additional strain on the landing gear.
3. Affords best directional control, especially at the beginning of the run.
4. Results in better obstacle clearance owing both to a shorter run and a steeper angle of climb (Fig. 2-36).
5. Establishes circuit pattern direction for all aircraft in the case of an uncontrolled airport.

The safety, and very frequently the quality, of a take-off can depend on the proper execution of the pre-take-off check. You must consider carrying out this check in the sequential manner prescribed by the authority responsible for the operation of the aircraft, or as laid down in the Aircraft Flight Manual, as a compulsory action at all times.

The first few yards (metres) of any take-off are very important. If a good, straight, well-controlled start is made, the success of the take-off is fairly well assured. Avoid the use of brakes if possible, since any use of brake will cause an undesirable increase in the take-off distance.

At first it is difficult to appreciate the varying control pressures required as the speed of an aircraft

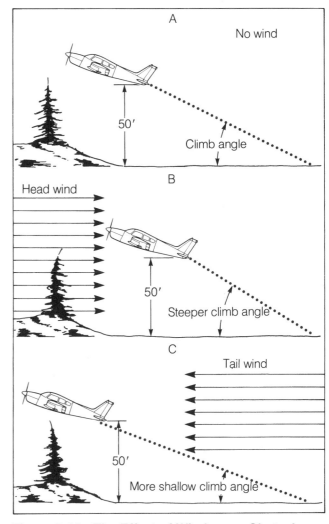

Figure 2-36 The Effect of Wind on an Obstacle Clearance Climb

increases during the take-off run. Therefore, there is a tendency to move the controls through a wide range, seeking the pressures expected, and as a consequence, to overcontrol badly. This will be aggravated by the initial sluggish reaction of the aircraft to the control movements. It is necessary that you develop a feel for control resistance and accomplish the desired results by pressing against them. With increased practice and experience, you will be able to sense also when sufficient speed has been attained for rotation so that you do not have to direct your attention to the airspeed indicator too soon. Achieving the recommended airspeed before beginning rotation is important, but until that time you must give full attention to outside references.

Nose Wheel Aircraft.   For normal take-offs in nose wheel aircraft, the aircraft should be carefully aligned with the runway centre line. Ensure that the nose wheel is centred. Power should be applied by opening the throttle smoothly but positively. Keep the ailerons and elevator in the neutral position. As the take-off roll commences, gradually move the elevator control back to lighten the weight on the nose wheel. As the speed of the aircraft approaches that required for take-off, raise the nose to the take-off attitude. Premature or excessive raising of the nose will delay take-off because of the increased drag. Keep straight by concentrating on a reference point at the far end of the runway and maintain directional control with smooth rudder pressures. Keep the wings level with aileron control.

Nose wheel aircraft are not ordinarily subject to gyroscopic effect or asymmetric thrust during the take-off run since, until rotation for lift-off is executed, the thrust-line remains constantly parallel to the ground. The exception would be the nose-up attitude required in a soft field take-off.

When the aircraft lifts off the ground, it should be at approximately the attitude for its best rate of climb airspeed and allowed to accelerate to this airspeed before any attempt is made to reset the throttle for the climb. Some pressure is normally required on the elevator control to hold this attitude until the proper climb speed is established. If back pressure on the control column is relaxed before a climb has been established, the aircraft may "settle," even to the point of dropping down onto the runway again.

The best rate of climb speed should be maintained until a safe height is reached, except when there are obstacles in the flight path, in which case the best angle of climb speed should be used. Unless the Aircraft Flight Manual states otherwise, light aircraft should maintain full power until at least 500 feet above the ground. The combination of full power and best rate of climb speed gives an additional margin of safety in that altitude is gained, in a minimum of time, from which the aircraft can be safely manoeuvred should there be an engine failure. Also, in many light aircraft, full throttle provides a richer mixture for additional engine cooling during the climb.

Tail Wheel Aircraft.   At the outset of the take-off run, keep the tail wheel on the ground so that directional control may be maintained by means of tail wheel steering. When speed has reached the point where both the elevators and rudder become effective, lower the nose to the take-off attitude.

As take-off speed is attained, ease the control column back to assist the aircraft into the air. It should not be forced into the air prematurely.

There may be a tendency for the aircraft to yaw to the left during the take-off run. The main reason for this is undoubtedly the propeller, but which of its effects is the chief cause is not so easy to determine. For example:

1. The torque reaction of the clockwise rotating propeller being anti-clockwise, the left main wheel will be pressed on the ground and the extra friction at that point should tend to yaw the aircraft to the left. (This is similar to having a partially flat tire on the left front wheel of an automobile.)
2. The slipstream will strike the fin and rudder on the left side, tending to yaw the aircraft to the left.
3. Gyroscopic effect will enter the picture when the tail is being raised, and will tend to yaw the aircraft to the left.
4. Asymmetric thrust. In the case of a tail wheel aircraft, since the right side of the propeller's plane of rotation is developing more thrust than the left side, it will cause the aircraft to yaw to the left, while the tail wheel is on the ground.

Cross-Wind Take-offs

The strength and direction of the wind is an important consideration for take-offs. It is not always practical to take off directly into the wind. As a result, cross-wind take-offs must be practised until the skill has been mastered.

For all take-offs, the aircraft must maintain a straight path along the centre of the runway. During a cross-wind take-off, the wind blowing across the runway makes directional control more difficult and must be counteracted. There is a tendency for the into wind wing to rise causing the aircraft to roll. Wind moving across the runway may also make the aircraft turn into the wind or drift sideways.

At least four factors account for the aircraft's reactions to a cross-wind (Fig. 2-37):

1. The aircraft has more keel surface behind the main wheels than ahead of them. As a result, the wind exerts a greater sideways force on the rear portion

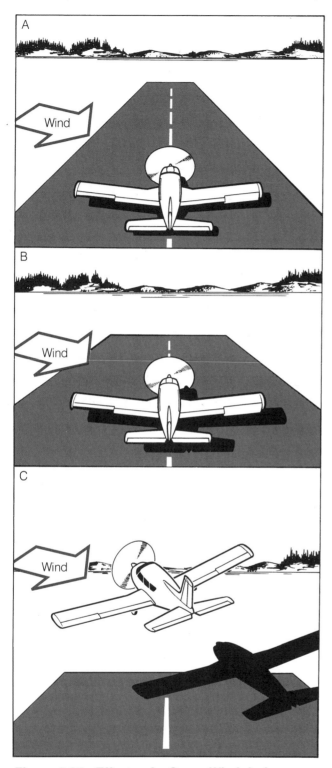

Figure 2-37   Effects of a Cross-Wind during Take-off

3. The into wind wing is exposed to more wind because the fuselage shelters the other wing somewhat. The wing that receives the greater effect from the wind will produce more lift.
4. With positive dihedral, the angle of attack on the into wind wing is larger than the angle of attack on the other wing. Again, the wing that receives the greater effect from the wind will produce more lift than the other wing.

During a cross-wind take-off, directional control is maintained with rudder. Depending on the strength of the wind and the aircraft type, much greater rudder pressure than normal may be required to maintain a straight path along the runway.

During a cross-wind take-off, ailerons are deflected as though turning into the wind. This counteracts the tendency for the into wind wing to rise. Full aileron should be deflected as the take-off roll begins. As the aircraft gains speed along the runway, aileron deflection will be reduced. Whenever a cross-wind is present, some into wind aileron deflection will be required at the point of lift-off to counteract the tendency for the aircraft to roll.

The aircraft must leave the runway surface cleanly and positively. Once the aircraft is airborne, it must not be allowed to settle back onto the runway. If the aircraft were to drift sideways after take-off and settle back onto the runway, the sideways movement would create a considerable side loading on the landing gear. Should the runway be quite narrow, the aeroplane could drift completely off the runway onto the unprepared surface along the side.

As soon as the aircraft is airborne and there is no possibility of settling back on the runway, a co-ordinated turn is made into wind. The turn is stopped and the wings levelled when the new heading compensates adequately for drift. A steady climb is maintained on this heading, which should result in a ground track aligned with the centre line of the runway. From time to time the term *crab* or *crabbing* may be used to describe an alteration to the heading of an aircraft to compensate for drift. Crabbing is a very descriptive and convenient term, but always bear in mind that the activity being described is relative to the ground only.

Except in the case of a direct head wind or a 90 degree cross-wind, a wind from any forward angle contains both a cross-wind component and a head-wind component. A wind blowing at a 90 degree angle contains only a cross-wind component. One of the simpler methods for determining acceptability of crosswinds uses this principle as a basis for calculation.

It is a certification requirement that an aircraft be capable of safe operation in a 90 degree cross-wind provided the speed of the wind does not exceed 20

of the aeroplane which causes the nose to turn into wind. This is referred to as "weather cocking."
2. The wind blowing across the runway tends to push the entire aircraft sideways creating a sideways strain on the landing gear.

percent of the stalling speed of the aircraft in question. This information, in conjunction with the known stalling speed of a particular aircraft, makes it possible to use the cross-wind component graph (Fig. 2-38) to derive a general rule for most light aircraft. This method must be used as a guide only, since acceptability of winds of any angle or strength depends on all circumstances involved, including the pilot-in-command's level of competence. Examples of the method used in this interpolation are shown below:

Example 1 Aircraft with a CAS Stalling Speed of 60 KT

Wind (Degrees off Runway)	Permissible Wind Speeds
90 degrees (0.2 × 60 KT stalling speed)	= 12 KT
60 degrees Using cross-wind component graph	= 14 KT
30 degrees Using cross-wind component graph	= 24 KT
15 degrees Using cross-wind component graph	= 45 KT

Example 2 Aircraft with a CAS Stalling Speed of 50 KT

Wind (Degrees off Runway)	Permissible Wind Speeds
90 degrees (0.2 × 50 KT stalling speed)	= 10 KT
60 degrees Using cross-wind component graph	= 12 KT
30 degrees Using cross-wind component graph	= 20 KT
15 degrees Using cross-wind component graph	= 38 KT

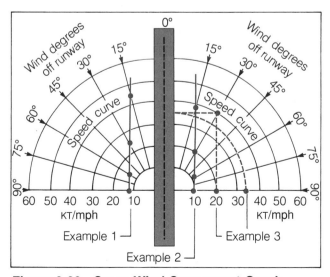

Figure 2-38 Cross-Wind Component Graph

When you have determined that a strong cross-wind is within acceptable limits, it is often important to know the value of the head-wind and cross-wind components. Both of these values may be determined by using the graph in Fig. 2-38. Say, as an exaggerated example, the wind is 30 degrees off the runway at 40 KT (Example 3). The point where the 40 KT "speed curve" intersects the 30 degree "wind degrees off runway" line becomes the datum. Draw a vertical line down from the datum, and where it intersects the "KT/mph" line read off the cross-wind component (20 KT). Draw a horizontal line from the datum to the "runway edge," then from this point parallel the "speed curves" to a point on the "KT/mph" line and read off the head wind component (34 KT).

Special Considerations

Most flight training occurs at aerodromes with firm surfaces that are of adequate length and free from obstacles, which allows for normal take-offs. However, many take-offs must be made under less than ideal conditions. You may have to carry out take-offs from locations with:

1. Reduced overall runway length.
2. Snowdrifts, puddles of water, or other hazards limiting available runway length.
3. Obstacles (Fig. 2-39).
4. Soft or rough surfaces (Fig. 2-40).

Whenever special take-off conditions exist, consult the Aircraft Flight Manual for essential information. The manual may provide special procedures to follow or indicate the configuration that will provide optimum performance as well as graphs, charts, or other methods to determine take-off performance such as distance required to lift off or the distance required to climb over obstacles. When you calculate the take-off and climb distance required, you may decide against a take-off until more favourable conditions exist.

Take-offs From Firm Surfaces With Reduced Field Length

To maximize aircraft performance in these situations you will need to configure the aircraft to allow a safe lift-off at a slow speed. You will also have to accelerate as quickly as possible to this lift-off speed. The Aircraft Flight Manual will help you determine how to proceed.

Depending on the aircraft being used, the configuration that allows the slowest safe lift-off speed may involve flaps up or extended only partly. Consult the Aircraft Flight Manual to determine the flap setting for your aircraft.

Maximum acceleration is accomplished by ensuring before the take-off begins that maximum power is being produced. With brakes applied on a firm surface, the maximum power you can obtain should be compared with figures provided by the aircraft manufacturer. Failure to obtain the power readings specified in the manual is an indication that the engine is not delivering maximum power. This means that take-off performance will not be what you expect, therefore, you should not attempt the take-off. If full power is available, you should begin the take-off roll with full power. Most manufacturers suggest that, when possible, you should apply power against the brakes until the engine is developing full power and then release the brakes.

Normally, the take-off roll is completed with the aircraft in the attitude that provides the least drag. This attitude is similar to the normal cruise attitude. Allow the aircraft to reach the correct lift-off speed before raising the nose for the take-off. Lifting off the ground at too slow a speed will result in control difficulties or even a stall.

Once airborne, allow the aircraft to accelerate to the desired speed and climb away.

Take-offs From Soft or Rough Surfaces

Soft and rough surfaces require more distance for an aircraft to accelerate to flying speed. Take-offs from soft or rough surfaces are normally accomplished by completing all required check-lists on whatever solid surface is available and then taxiing and taking off without stopping on the soft or rough surface. Stopping may mean getting stuck and will require large amounts of power to get the aircraft moving. The aircraft is configured in accordance with the Aircraft Flight Manual.

A nose-high attitude is maintained during the take-off roll. This procedure will result in the aircraft becoming airborne at a slower speed than that associated with normal take-offs. This speed will be slower than required for a safe climb. In some cases, due to ground effect, the lift-off speed may be close to the speed normally associated with a stall. Immediately after lift-off, the aircraft must be transitioned into level flight with the wheels just clear of the surface until the desired climb speed is achieved. Care must be taken to prevent climbing until the proper climb speed is attained. Care must also be taken to

ensure that the aircraft is not forced back onto the runway.

Aside from maximizing acceleration, any steps that will reduce minimum flying speed can lead to reduced take-off distance. In some light aircraft, the use of small flap extensions will allow a safe lift-off at a lower than normal speed. The use of large flap extensions generally causes excessive amounts of drag. Check the Aircraft Flight Manual to determine the flap settings recommended for take-off.

Tricycle gear aircraft present an additional impediment to take-offs from soft surfaces. The nose gear may dig into the soft surface unless elevator forces are used to raise the nose wheel above the surface. While this increase in angle of attack causes additional drag, the drag is less than the resistance encountered by the nose wheel on a soft runway surface.

Obstacle Clearance Take-offs

Whether taking off from a solid surface or from a soft or rough one, it is possible that a further take-off problem exists in the form of obstacles over which the aircraft must climb. Whenever obstacles are of concern, the take-off must begin with the procedure dictated by the surface conditions. Soft fields still require soft field techniques; short fields still require short field techniques.

Once airborne, the aircraft must be operated so as to provide the greatest gain in height when compared to the distance travelled over the ground. When taking off and climbing over an obstacle the best angle of climb speed (Vx), as provided in the Aircraft Flight Manual, should be used. After lift-off at the recommended speed, the aircraft should accelerate to the best angle of climb speed and should be transitioned into a climb at this speed. This speed should be maintained until the obstacle has been cleared or until it is obvious that a normal climb would be appropriate. When the top of the obstacle appears to be below the level of the horizon, the aircraft is higher than the obstacle. At this point a normal climb should be established to improve forward visibility and safety.

For some aircraft, partial flap settings are recommended to reduce take-off distance. In some cases, the use of partial flaps improves climb performance, in other cases it reduces climb performance. Check the Aircraft Flight Manual for the aircraft you are flying to determine the procedure and flap setting recommended.

Aircraft weight affects climb performance. Many aircraft use slightly different speeds for the climb over an obstacle, depending upon the aircraft weight.

Again, consult the Aircraft Flight Manual for the aircraft you are flying.

Tail Wind After Take-off

Due to surface friction and other causes, it is possible to have a condition of "no wind" at ground level, but at several feet above the ground, sufficient wind to affect climb performance. Should this phenomenon develop into a tail wind aloft, there is a risk of the aircraft not being able to clear obstacles adequately in the climb out. When the wind is calm or light and variable at the surface, carry out a take-off procedure that makes adequate allowance for the possibility of a tail wind shortly after the aircraft has left the ground.

Hydroplaning

The wheels of an aircraft rolling on a wet paved runway press a "bow wave" ahead of them. This can cause a film of water to develop between the tires and the runway of sufficient strength to "float" an aircraft during the take-off run. Under these conditions the aircraft may drift sideways and brakes can become ineffective, making control of the aircraft difficult at critical points during the take-off. When raindrops appear to bounce on the runway, the possibility of hydroplaning should be suspected. Depressions in the runway that cause extensive pooling to occur during heavy rain or spring thaws may also cause an aircraft to hydroplane during the take-off run. If a take-off must be made under suspected hydroplaning conditions, be prepared to control the aircraft without the aid of brakes after an estimated ground speed of 30 KT has been achieved.

Ground Effect

Anything that will impede the acceleration of the aircraft during its take-off run, such as mud, slush, snow, surface irregularities, grass (or other vegetation), grade, etc., must be fully considered in respect to the take-off distance penalty that these factors may impose. Under these circumstances it would be wise to use the soft field take-off technique, which makes use of a phenomenon called *ground effect*. This is due to the effect of the ground on the airflow patterns about a wing in flight. Ground effect results in decreased induced drag; thus, making it possible for

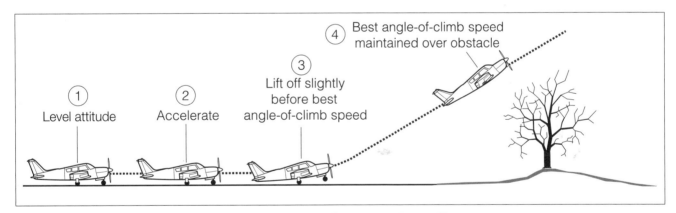

Figure 2-39 Rotation and Lift-off During Obstacle Clearance Take-off

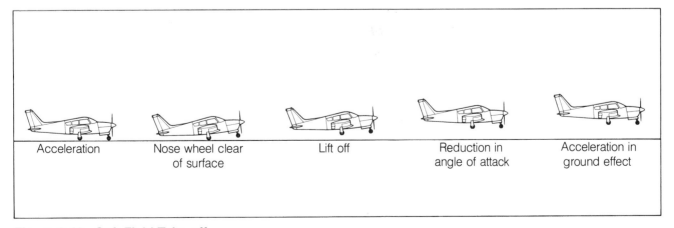

Figure 2-40 Soft Field Take-off

an aircraft to become airborne at less than normal airspeeds. As a general rule the results of ground effect can be detected up to a height equal to one wing span (of the aircraft being used) above the surface (Fig. 2-41). The phenomenon of ground effect has two important aspects which, if not recognized, can be extremely hazardous. Any attempt to climb out of ground effect prior to reaching the best angle of climb airspeed may result in the aircraft settling back to the surface of the runway. Secondly, it is possible to lift an aircraft off the ground into ground effect with insufficient power or too great a load to permit the aircraft to climb out of ground effect.

Performance

Many of the finer points associated with other than normal take-offs vary as to aircraft type and existing conditions. Aspects of these procedures are continually open to discussion and the apparent solutions will vary with the background and experience of the individual pilot. Such points include whether to apply take-off power before releasing brakes or as the aircraft moves forward, when commencing a take-off; whether to use flaps or not to use flaps; whether to extend flaps before the take-off run commences or as the take-off roll progresses. The answers to these questions are generally outlined in the procedures recommended in the Aircraft Flight Manual, which should be carefully followed. Other points of judgement arise as to the advisability of commencing to build up speed for take-off while moving from the run-up position to the runway centre line on a short field or obstacle take-off. The decision to follow this technique must be governed by the manufacturer's recommendations. Some types of aircraft may suffer an engine failure on the take-off roll following the so-called "rolling take-off" because the fuel will flow to one side of the tank, due to centrifugal force, leaving the outlet momentarily dry. This exposes the engine to fuel starvation and the possibility of failure at a critical point of the take-off. The amount of fuel in the tanks is another governing factor.

During a soft field take-off the taxiways and proposed take-off surface must be carefully inspected for extra soft spots. Provided the surfaces are deemed usable, the inertia of the aircraft and perhaps extra power must be used to carry it through these areas.

The base measurement for development of the performance data for a particular aircraft is customarily the performance achieved by the aircraft in a standard atmosphere, or as it is sometimes called, standard air density. The standard atmosphere is the air density when the barometric pressure is 29.92 inches of mercury and the temperature is 15°C (59°F).

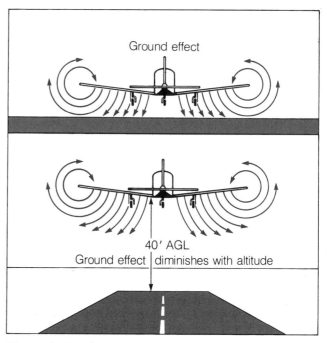

Figure 2-41   Ground Effect

The density of the air plays an important part in the take-off performance of an aircraft. Cold, dry air is denser than hot, moist air, and the denser the air, the better the performance. Factors to remember about air density at airports are:

1. Airport elevation high, air less dense = reduced performance.
2. Ambient air temperature high, air less dense = reduced performance.
3. Relative humidity high, air less dense = reduced performance.
4. Combination of 1, 2, and 3 = poor performance (Fig. 2-42).

Good take-off performance can be expected from an aircraft at an airport with a field elevation, say, of 150 feet above mean sea level on a cold day, whereas a poorer performance can be expected from the same aircraft at the same airport on a hot day.

The worst possible take-off (and climb) performance can be expected when the following four conditions are combined:

1. Air temperature — High (above 15°C).
2. Airport elevation — High.
3. Atmospheric pressure — Low (below 29.92).
4. Relative humidity — High.

(The above combination represents a high density altitude.)

Density altitude is the altitude corresponding to a given density in a standard atmosphere. It is a

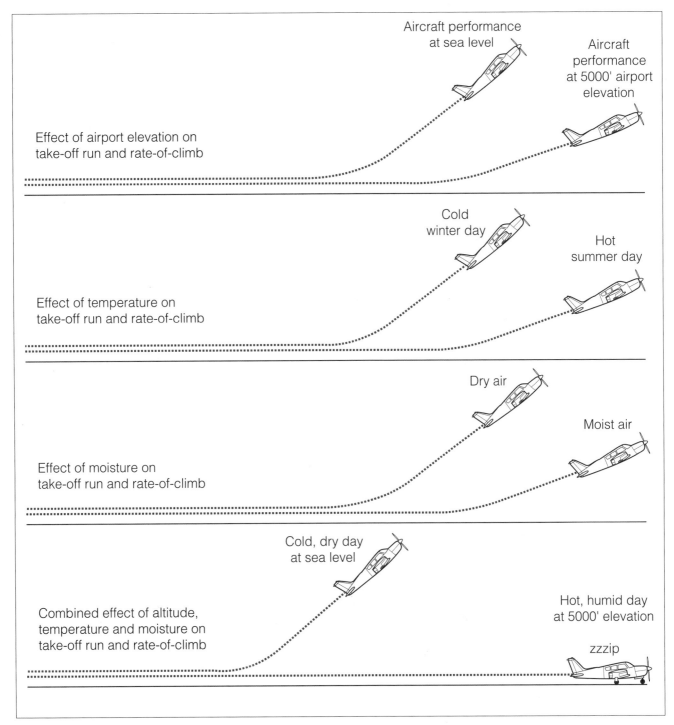

Figure 2-42   Effects of Elevation, Temperature, and Moisture on Take-off Run and Rate of Climb

"condition," not a level of flight. Unless density altitude is known it is difficult to determine the performance of an aircraft accurately, and this can be a very important factor under certain take-off conditions. Density altitude can be calculated very quickly on the pilot's flight computer.

Full use should be made of the take-off distance tables in the Aircraft Flight Manual, which show the changes in performance resulting from various airport elevations and air temperatures. The aircraft manufacturer's recommendations are always the best source for this information but should these recommendations not be available, useful take-off performance data may be calculated by using the Koch chart (Fig. 2-43).

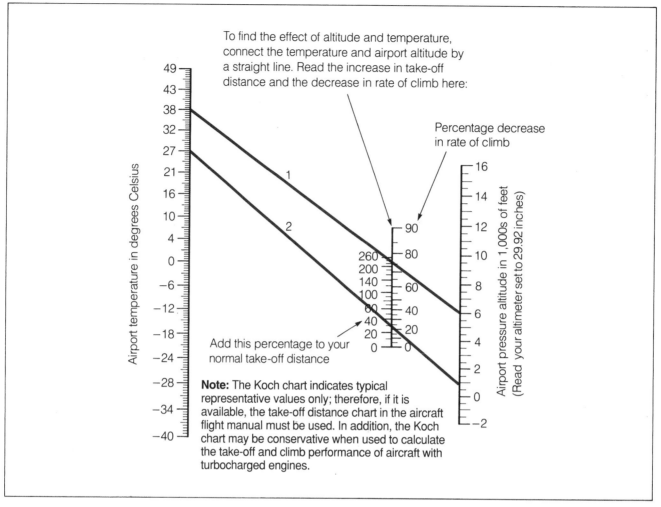

To find the effect of altitude and temperature, connect the temperature and airport altitude by a straight line. Read the increase in take-off distance and the decrease in rate of climb here:

Percentage decrease in rate of climb

Airport temperature in degrees Celsius

Airport pressure altitude in 1,000s of feet (Read your altimeter set to 29.92 inches)

Add this percentage to your normal take-off distance

Note: The Koch chart indicates typical representative values only; therefore, if it is available, the take-off distance chart in the aircraft flight manual must be used. In addition, the Koch chart may be conservative when used to calculate the take-off and climb performance of aircraft with turbocharged engines.

Figure 2-43   The Koch Chart for Temperature and Altitude Effects

Koch Chart

When using the Koch chart remember that the airport altitude factor is *pressure altitude*. To determine the pressure altitude of an airport upon which the aircraft is standing, set the altimeter barometric pressure scale to 29.92, then read off the altitude in the normal manner. After determining the pressure altitude of the airport, do not forget to reset the altimeter to the actual field elevation or altimeter setting.

The straight line (1) used as an example in the Koch chart (Fig. 2-43) shows that an aircraft at an airport with a pressure altitude of 6,000 feet with an outside (ambient) air temperature of 38°C (100°F) requires 220 percent more take-off distance than the same aircraft would require at sea level in standard atmosphere. Even at a common airport pressure altitude of 1,000 feet, 30 percent more take-off distance is required if the ambient air temperature is 27°C (80°F) (line 2).

The Koch chart indicates representative values only. Therefore, if it is available the take-off distance chart

in the Aircraft Flight Manual must be used. In addition, the Koch chart may be conservative when used to calculate the take-off and climb performance of aircraft with turbocharged engines.

Wheelbarrowing

Wheelbarrowing may be described as a condition in nose wheel equipped aircraft that is encountered when the nose wheel is firmly on the runway and the main wheels are lightly loaded or clear of the runway during take-off or landing. This causes the nose gear to support a percentage of weight greater than normal while providing the only means of steering.

During take-off, wheelbarrowing may occur at relatively low speeds due to the slipstream increasing the lifting effect of the horizontal stabilizer, and excessive forward elevator control pressure being applied during take-off to hold the aircraft on the ground to speeds above those normal for take-off. When taking off in a cross-wind, or if any other yaw force is introduced at this time, an aircraft in this flight condition tends to pivot about the nose wheel, and if not brought under control quickly may execute a manoeuvre similar to a ground loop in a tail wheel type aircraft.

Corrective action must be based on a number of factors — degree of development of the wheelbarrowing, the pilot's proficiency, remaining runway length, and aircraft performance versus aircraft configuration. Only after considering at least these factors should you initiate one of the following corrective measures:

1. If the aircraft is not pivoting, ease back on the control column to take weight off the nose wheel and continue with the take-off and climb procedure.
2. If pivoting has begun, relax forward elevator control to lighten the load on the nose wheel and return steering to normal. If pivoting stops, resume the take-off; if pivoting continues, reject the take-off.

Wake Turbulence

Wake turbulence (Fig. 2-44) caused by wing tip vortices of departing or arriving aircraft, especially large, heavy aircraft, must be avoided by aircraft about to

take off. An aircraft flying into the core of a wing tip vortex will tend to roll with that vortex. It is entirely possible that this induced roll will be at a greater rate than a light aircraft's capability to counteract it.

Wake turbulence from any preceding aircraft will be maximum just before the point of touchdown for a landing aircraft, and just after the point of take-off for a departing aircraft. On take-off, wake turbulence can best be avoided:

1. If following an aircraft that has just departed, by planning the take-off so as to become airborne prior to the point of take-off of the preceding aircraft. Avoid passing through the flight path of the preceding aircraft.
2. If following an aircraft that has just landed, by planning the take-off so as to become airborne beyond the point of touchdown of the preceding aircraft (Fig. 2-45).

Should you have the slightest doubt or indecision concerning wake turbulence on take-off, delay the take-off for up to two minutes to allow the vortices of a landing aircraft to dissipate and up to four minutes in the case of a preceding take-off. The larger the preceding aircraft, the longer the delay. Remember that

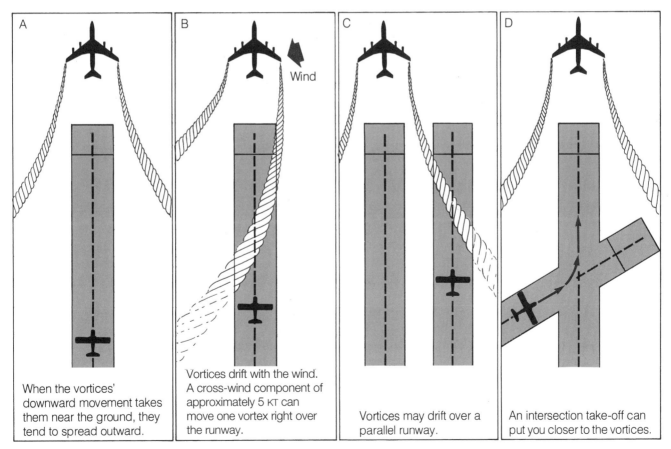

Figure 2-44 Wake Turbulence

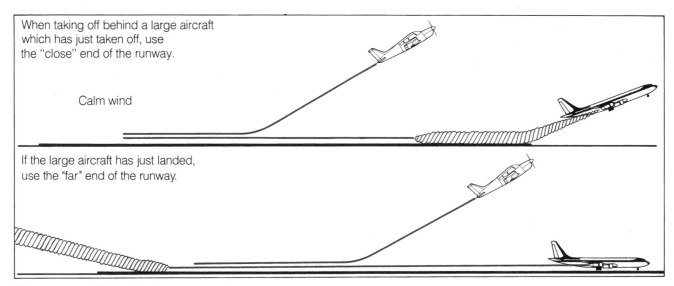

When taking off behind a large aircraft which has just taken off, use the "close" end of the runway.

Calm wind

If the large aircraft has just landed, use the "far" end of the runway.

Figure 2-45   Wake Turbulence Avoidance Procedures

although the strength of the wind is a vortex dissipation factor, a cross-wind may move or hold a vortex directly in the proposed take-off path.

It is unlikely that a control tower would clear a light aircraft for an immediate take-off in the wake of a large heavy aircraft, but in any case there should be no hesitation whatsoever on the part of the light aircraft pilot in requesting a take-off delay, should such a clearance be given.

Even though a clearance for take-off has been issued, if you consider it safer to wait or alter your intended operation in any way in the interest of safety or good airmanship, ask the control tower for a revised clearance. The air traffic controller's chief interest is safety of aviation, but the controller may not be aware of all circumstances, especially a pilot's level of competency when unusual conditions prevail.

At an uncontrolled airport (an airport without a control tower), the pilot-in-command is responsible for making decisions that are consistent with safety, good airmanship, and the *Canadian Aviation Regulations*. In this regard, two of the most important *Canadian Aviation Regulations* indicate that:

1. Where an aircraft is in flight or manoeuvring on the ground or water, the pilot-in-command shall give way to other aircraft landing or about to land.

A landing aircraft has priority in the use of the landing area and an aircraft proposing take-off must not usurp this priority in any way. Whether a take-off may be made, in view of the distance a landing aircraft is from the landing area, is a matter of good judgement and courtesy. Never presume that a landing aircraft can always abort its landing in favour of a pilot who has exercised poor judgement in timing the take-off.

2. No person shall take off or attempt to take off in an aircraft until such time as there is no apparent risk of collision with any other aircraft.

Search the entire sky for other aircraft that may conflict with your take-off. Another aircraft may be landing with a tailwind on the runway; this may be in conflict with the *Canadian Aviation Regulations* but such an aircraft still has priority over aircraft proposing take-off. The aircraft landing with a tailwind may have reasons compelling it to do so. Do not take off, or position your aircraft for take-off, until a landing aircraft has cleared the runway. When positioned for take-off, the pilot no longer has a view of the runway approach and possible landing traffic.

It is an indication of poor airmanship and lack of courtesy on your part to proceed onto the active runway if you are not fully prepared to take off as soon as the runway is clear. Remain clear of the active runway until you have carefully completed all pre-take-off checks. Do not delay aircraft taxiing behind you by carrying out cockpit procedures that should have been completed at the apron or ramp. If there is some difficulty that may delay your take-off, position your aircraft so that others may pass.

The Circuit

The International Civil Aviation Organization (ICAO) terminology for the circuit is "aerodrome traffic circuit." It is defined as: "The specified paths to be flown by aircraft operating in the vicinity of an aerodrome." The circuit is often erroneously referred to as the "traffic pattern." Although the latter does involve the circuit, the correct definition of traffic pattern is: "The geographical path flown by an aircraft after it enters a control zone and until it enters the downwind leg of the aerodrome traffic circuit."

The prime purpose of an orderly and well defined circuit is safety. However, circuit procedures are also fundamental to the execution of good approaches and landings.

The basic pattern of the circuit remains fixed, but its orientation is determined by the heading of the runway in use at the time. A plan view of the circuit (Fig. 2-46) shows that it is rectangular in shape and has the following components:

1. Take-off.
2. The cross-wind leg (not to be confused with circuit joining cross-wind).
3. The downwind leg.
4. The base leg.
5. The final approach.

In actual practice, at controlled airports it is customary for pilots and controllers to omit the word "leg" when referring to the circuit components, e.g.: "Burton tower / FOXTROT, ALPHA, BRAVO, CHARLIE / downwind;" "FOXTROT, ALPHA, BRAVO, CHARLIE / Burton tower / report turning base." It is recommended that the downwind call be when the aircraft is abeam the control tower.

Unless special conditions exist and there is authorized advice to the contrary, all circuits are left hand; therefore, all turns within the circuit are left turns.

In addition, unless otherwise authorized, all normal circuit heights are 1,000 feet above aerodrome elevation.

After take-off there will be a straight climb into wind, normally to a height of 500 feet, and then a 90 degree turn cross-wind. The cross-wind leg is a continuous climb to circuit height where the aircraft is levelled off. Then a 90 degree turn brings the aircraft onto the downwind leg. The downwind leg is flown so as to track parallel with the intended landing path. On the downwind leg any necessary pre-landing checks are made. When past the downwind boundary an appropriate distance, another 90 degree turn is made onto the base leg. When within gliding distance of the landing area and a gliding approach is desired the throttle is closed and the aircraft is put into a glide. Just before reaching the intended line of the final approach, another 90 degree turn is made onto final approach and the aircraft is kept in line with the centre of the runway until the landing is completed (see Fig. 2-46).

The strength of the wind will affect the heading to be steered on the cross-wind and base legs of the circuit. Maintain a track over the ground, which is at right angles to the landing path. Thus, during the base leg, an aircraft in a very strong wind will be heading well in toward the aerodrome although its path over the ground will be at right angles to the landing path.

On all legs of the circuit it is essential to maintain a good look-out on both sides, above, and below.

The latter part of the circuit is usually called the approach. Technically, the approach to landing commences on the downwind leg, at the turning point from downwind to base leg. Exactly where or when the turn onto the base leg is made will depend largely on the strength of the wind. The stronger the wind, the steeper the angle of descent will be during the final straight approach; therefore, the sooner the turn onto the base leg should be made.

When on the base leg, adjust the heading to allow for drift and judge when to start the glide (or descent) by the angle at which the runway is observed — the

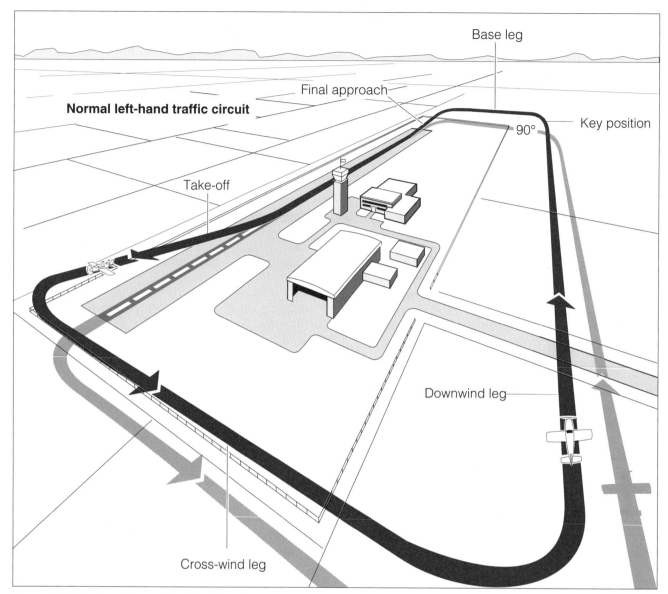

Figure 2-46   Aerodrome Traffic Circuit

stronger the wind the greater this angle should be. Once the descent has begun, you can judge whether the aircraft is going to be too high or too low on the final approach and correct by appropriate use of flaps or power.

Judge the whole circuit in relation to the runway, not in relation to other points on the ground. This will rapidly improve your judgement of approaches and changes of runway, landing direction, or aerodrome will not be upsetting. It will also help you considerably in judging other types of approaches.

Right Hand Circuits

The standard direction of any aerodrome traffic circuit is left hand. However, exceptions occur where traffic conflicts with other airports, or hazardous terrain

necessitate the adoption of a right-hand pattern, for an entire airport or for specific runways. The exceptions are listed in the *Canada Flight Supplement*.

Spacing

It is extremely important that you be constantly aware of the position of other aircraft in the circuit, particularly those that are ahead of you in the pattern. Be careful not to "cut off" a preceding aircraft by turning onto the base or final leg out of proper sequence. Maintain suitable spacing between your aircraft and the one ahead to allow that aircraft time to land and taxi clear of the runway. If you crowd the preceding aircraft it may be necessary for you to execute a missed approach and "go around," which in these circumstances is an unnecessary waste of flight time. At the same time, overspacing in a busy circuit will inconvenience aircraft that are following you. Correct spacing is a judgement skill you must develop as quickly as possible. It takes into account such matters as wind direction and strength, and the circuit speeds of other aircraft. Correct spacing may be accomplished by widening or narrowing your circuit and/or increasing or decreasing airspeed.

Controlled Airports

A *control zone* is controlled airspace about an airport or military aerodrome of defined dimensions extending upward from the surface to a specified height above ground level. Civilian control zones with control towers are normally designated as Class "B," "C" and "D," within which special regulations apply. Control zones without operating control towers are usually designated Class "E." For information concerning airspace classification, refer to the RAC section of the *Aeronautical Information Publication* (A.I.P.) Canada.

Leaving the Circuit — Controlled Airports

When an aircraft leaves the circuit after take-off, it does one of two things. It either operates outside the circuit while remaining within the control zone, or it leaves the control zone. When an aircraft remains within the control zone, the control tower will most likely ask that it: (1) remain on the control tower frequency; (2) advise the type of exercise; (3) advise the altitude at which the aircraft will be flown; and (4) advise where the flying will be carried out. When an aircraft intends to leave the control zone, permission to cease monitoring the control tower frequency must be granted by the control tower so long as the aircraft is in that tower's control zone. The control tower exercises jurisdiction over all VFR traffic within its control zone. A VFR aircraft may not operate within a control zone without permission from the appropriate control tower, even though the aircraft may be using another airport within the control zone.

When leaving the circuit, if the aerodrome traffic circuit is left hand you may execute a right-hand turn after take-off only with permission from the control tower. If this permission cannot be granted, you must follow the tower controller's instructions until clear of the zone.

After leaving the circuit it may be necessary to fly through the control zone of another airport at which you do not intend to land. It is compulsory that you make radio contact with the control tower in this zone and remain under its control until out of the zone again. Unless some special prior arrangement is made with the appropriate Air Traffic Control unit, aircraft

without two-way radio communication should remain clear of control zones.

Automatic Terminal Information Service (ATIS) is provided at many larger airports. The recorded broadcast includes weather, runway, and NOTAM information affecting the airport. Where it is provided, you should monitor the ATIS broadcast prior to calling the ATC facility, and inform ATC on first contact that you have received the pertinent information.

Joining the Circuit

When returning to an airport for landing, advise the control tower of your identification, geographical location, or estimated distance in miles and direction from the airport, and altitude. Then request landing instructions. If you are outside the control zone you must do this prior to entering the zone. When the control tower gives you clearance "to the circuit" you are expected to join the circuit on the downwind leg at circuit height. The descent to circuit height must be made outside of the area occupied by the circuit.

"Cleared to the circuit" authorizes you to make a right turn, if required, to join cross-wind, or to join the downwind leg provided the right turn is only a partial turn that can be carried out safely (Fig. 2-47). When cleared by the control tower for a "straight in" approach, you are authorized to join the circuit on the final approach leg without having executed any other part of the circuit. The same ruling applies to being authorized to join on base leg.

When an aircraft has been cleared to land by a control tower it does not mean the runway is clear of all hazards. Any known hazards will be mentioned at the time of the clearance; however, at times of restricted visibility or at night, when a view of the runway from the tower may be limited, unauthorized vehicles or animals may have moved onto the runway without the controller's knowledge. As the pilot, it is your final responsibility to keep a look-out for hazards on the landing and manoeuvring area. You alone must decide whether it is acceptable to land considering your aircraft and level of flying skill.

Uncontrolled Aerodromes

An uncontrolled aerodrome is an aerodrome without a control tower or one where the tower is not in operation. An airport which is a certified aerodrome (see Chapter 5) can also be uncontrolled. There may be no air-to-ground radio communication at an uncontrolled aerodrome. However, at many sites there is a mandatory or aerodrome traffic frequency on which communication can be established with a Flight Service

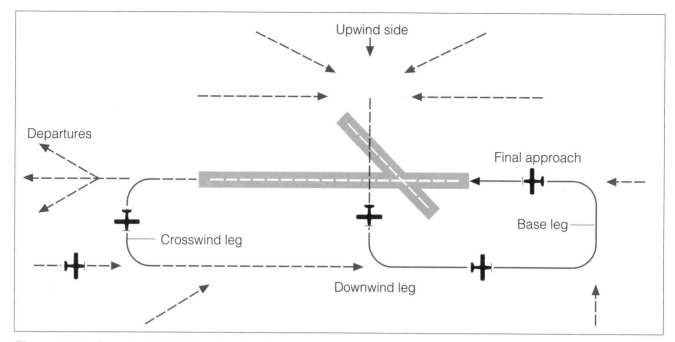

Figure 2-47   Standard Left-Hand Traffic Circuit at Controlled Airports

Station or a locally based aircraft operator. These facilities exercise no control over aircraft, but they can be very helpful in advising of surface winds, the runway being used by others, known air and ground traffic, runway conditions, weather, etc. Pilots are encouraged, and in many cases, required to monitor and make use of any radio or unicom communications that may be available at uncontrolled aerodromes, and to transmit position reports and broadcast their intentions before joining and while in the circuit.

Leaving the Circuit

After take-off, climb straight ahead on the runway heading until reaching the circuit traffic altitude before commencing a turn in any direction to an en route heading. Turns back toward the circuit or airport should not be initiated until at least 500 feet above the circuit altitude.

Joining the Circuit

When returning to the airport for landing take full advantage of air-to-ground communications for advice. Many conditions can change at an airport after even a short absence. If you cross the airport to make observations, the cross-over must be done at least 500 feet above the circuit altitude. The descent to circuit height should be made on the upwind side so as to join the circuit at circuit altitude in level flight. Under normal circumstances, circuit height is 1,000 feet above aerodrome elevation.

Where no mandatory frequency procedures are in effect, aircraft should approach the traffic circuit from the upwind side, or, if no conflict exists with other traffic, the aircraft may join the circuit on the downwind leg (Fig. 2-48).

Where mandatory frequency procedures are in effect and airport and traffic advisory information is available, aircraft may join the circuit pattern straight in or at 45 degrees to the downwind leg, or straight in to the base or final approach legs. Be alert for other VFR aircraft entering the circuit at these positions and for Instrument Flight Rules (IFR) aircraft on straight in or circling approaches. The pilot of an aircraft inbound on an IFR or practice instrument approach may give a position report that you do not understand. Do not hesitate to request clarification in order to gain a clear understanding of that aircraft's position and intentions.

Normally the runway to use for landing is the one most nearly aligned into wind. However, the pilot has final authority; therefore, for the safe operation of the aircraft another runway may be used if the pilot deems it necessary.

Taxiing on a Runway in Use

It is sometimes necessary to turn 180 degrees and taxi back down the runway to position the aircraft for take-off, or after landing in order to clear the runway. Do this as quickly as possible consistent with safety. Remember, after landing, until the runway is cleared no other traffic has landing priority.

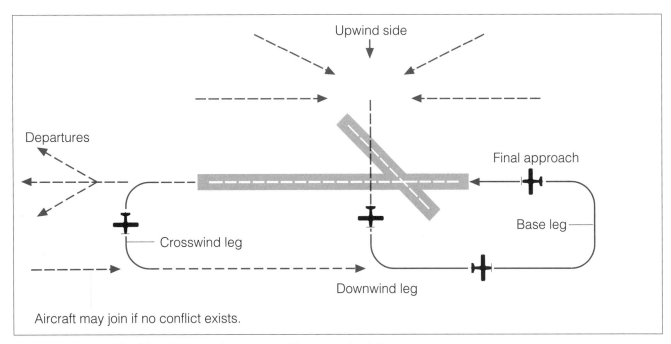

Upwind side

Departures

Final approach

Base leg

Crosswind leg

Downwind leg

Aircraft may join if no conflict exists.

Figure 2-48 Traffic Circuit Procedures at an Uncontrolled Airport

Approach and Landing

It is common for the student to believe that landing an aircraft is the sum total of flying, and that once this is learned, about all there is to know has been accomplished. This belief, if allowed to persist, produces two unfortunate results: first, mental hazards, based on attaching undue importance to the landing procedure, which may hinder progress in learning the procedure; and second, slacking off once reasonable proficiency in executing landings has been attained.

Actually the landing is just another manoeuvre, representing the logical result of all the preparation up to this point, and only one of a series of extensions of principles by which learning has progressed and will continue to progress toward the goal of pilot competency. A landing is the last of a sequence of major manoeuvres, during which the altitude must be controlled, air traffic observed, and the whole process performed safely with an acceptable degree of proficiency.

Landing an aircraft consists of permitting it to contact the ground at the lowest possible vertical speed, and under normal circumstances, at the lowest possible horizontal speed consistent with adequate control. The first step toward reducing the horizontal velocity relative to the ground is to land into the wind; the second step is to obtain the desired airspeed and attitude at the appropriate moment.

Although the approach to landing and the landing itself may be considered two separate manoeuvres, one is usually an integral part of the other. The success of a landing depends on the type of approach technique used to meet the operational requirements of a specific landing procedure. Landings may be classed as follows:

1. *Normal* landing.
2. *Cross-wind* landing.
3. *Short field* landing.
4. *Soft* and *unprepared field* landing.

Except under the most ideal conditions, even a normal landing involves some degree of cross-wind, and other landings may involve a combination of all three classes. For example, a short field landing will very likely involve the techniques required for cross-wind landings and those for landing over an obstacle. The techniques for each class of landing will be treated separately in the sequence shown, beginning with the normal landing.

Normal Landings

A normal landing is a slow transition from the normal glide attitude to the landing attitude. This transition is referred to as the flare, or the round-out. It is started approximately 15 to 30 feet above the ground, and progressively increased and continued as altitude is lost until, in tail wheel aircraft, the main landing gear and the tail wheel touch the ground simultaneously. Aircraft with tricycle landing gear should contact the runway on the main landing gear, with no weight on the nose wheel (Fig. 2-49).

Your body's sense of motion will not have developed enough at this stage to be of primary assistance in landings, although it will be a factor. Vision is therefore the most important sense you use and you will operate the controls in accordance with it. Reactions on the controls to prevent the aircraft from flying into the ground will be instinctive, but some reactions are likely to be wrong, particularly as to degree and often as to type as well.

Accurate estimation of distance and depth, besides being a matter of practice, depends on how clearly objects are seen. It requires that your vision be focused properly so that important objects stand out as clearly as possible. Speed blurs objects at close range; nearby objects seem to run together while objects farther away stand out clearly. At the time of landing you should focus ahead of the aircraft, at

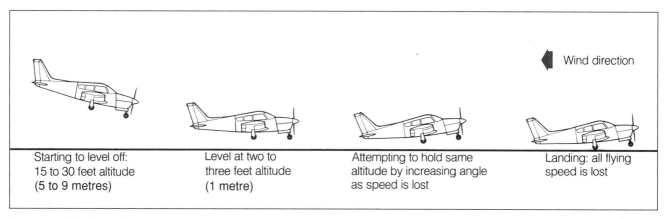

Starting to level off:
15 to 30 feet altitude
(5 to 9 metres)

Level at two to
three feet altitude
(1 metre)

Attempting to hold same
altitude by increasing angle
as speed is lost

Landing: all flying
speed is lost

Wind direction

Figure 2-49   Landing

about the same distance as you would in a car travelling at the same speed. The distance at which the vision is focused should be proportionate to the speed of the aircraft. Thus, as speed decreases, the distance ahead of the aircraft at which it is possible to focus sharply becomes closer; therefore, the focus should be brought closer accordingly. However, if your vision is focused too closely, or straight down, objects become blurred and reactions will be either too abrupt or delayed too long.

At the very outset, form the habit of keeping one hand on the throttle control throughout the landing. If a situation suddenly arises that requires an immediate application of power, the time necessary for recognizing the problem, moving the hand to the throttle, opening it, and having the engine respond, is too great. Bounces are common at the initial stages of training and proper use of the throttle at the exact instant is imperative.

In addition to practising power assisted approaches, at every opportunity you should practise landings from full glides, with the engine throttled back to idling. This type of approach is very necessary to develop the judgement and planning required for forced landing procedures.

When the aircraft is within 15 to 30 feet (5 to 9 metres) from the ground, the flare (round-out) should begin. Once started, it should progress continuously until the aircraft is on the ground. If your speed is correct, as back pressure is applied to the control column the aircraft will begin to lose speed and start to settle. As the ground "comes up," continue to ease the control column back. This movement of the elevator control is timed so that the slow, smooth, continuous, backward movement holds the aircraft just above the surface until the desired landing attitude is attained. Nose wheel aircraft should contact the ground on the main wheels first, with no weight on the nose wheel. In tail wheel aircraft all wheels should touch the ground simultaneously, with the elevator control all the way

back and the throttle closed. This requires the development of fine timing, technique, and judgement of height and distance.

The point at which the aircraft is flared makes all the difference to the subsequent landing. Much research has been done with a view to finding out how an experienced pilot judges this point. Here are some suggestions which may be helpful:

1. Try to judge the height of the aircraft above the landing surface using the height of known objects.
2. Try to judge that point at which the ground seems to be coming up so rapidly that something must be done about it.
3. Watch the ground where touchdown is expected. When it appears to start to approach rapidly, check the rate of descent by easing the control column back.
4. Note the point at which the whole area of the landing surface seems to expand.
5. Note the point at which movement of ground suddenly becomes apparent.

Once the actual process of landing is started, the elevators should not be pushed forward to offset any ordinary error in backward movement of the controls. If too much back pressure has been exerted, this pressure may be either slightly relaxed or held constant, depending on the degree of error. In some cases it may be necessary to advance the throttle slightly to compensate for a loss of speed.

When the aircraft has come to within 2 or 3 feet (1 metre) of the ground, check its descent by further back pressure on the elevator control. At this point the aircraft will be very close to its stalling speed; therefore, backward pressure does not increase or maintain height as might be expected. Instead, it slows up the settling phase, so that the aircraft will touch the ground gently in the desired landing attitude. Remember, as was evident in your slow flight training, as airspeed decreases more control movement will be needed to gain the desired effect.

The completion of the touchdown should be judged by the change in attitude of the aircraft rather than by movements of the control column. The attitude should be changed by reference to the landing horizon (edge of the aerodrome) and the front of the aircraft.

Once a tail wheel aircraft is on the ground, the control column should be held as far back as possible until the aircraft comes to a stop. This will shorten the landing roll and tend to prevent bouncing and skipping of the tail, together with improving directional control.

In the case of a nose wheel aircraft, allow the nose wheel to lower gently to the runway of its own accord as the forward speed decreases and the elevators lose their effectiveness. Do not relax your attention at this point. Keep straight. This type of aircraft should not normally be "flown on" and held on the runway with excessive speed, since this may impose excessive stress on the nose wheel and possibly cause the undesirable condition known as wheelbarrowing (discussed in Exercise 16).

As training progresses, you will be required to plan the approach and landing while the aircraft is on the downwind leg of the traffic circuit. This may be done by visualizing the flight path and estimating where you will reduce power and make the turn onto final approach to land. As practice progresses, the descent should be initiated by reducing the power and airspeed to produce the desired flight path on the base leg, and then making a 90 degree descending turn onto final approach. When the descent has been started, make drift corrections on the base leg to follow a ground track that will approximate a right angle to the runway. The base leg should be flown to the point where a gentle turn will bring the aircraft to the final approach directly in line with the landing runway. This turn must be completed at a safe altitude, which will depend on the elevation of the terrain at this point and the height of any obstructions. Make the final approach long enough to estimate the point of touchdown and allow for any necessary reductions of power and airspeed in preparation for the landing.

Flaps and Trim

To avoid undershooting a runway, there is often a natural tendency to be too high on the normal approach with the result that height must be lost. It is considered that the last 500 feet (152 metres) of a normal approach should be straight, without any slipping or turning, and that height should be controlled by the use of flap. Extending the flaps changes the airflow pattern over and under the wing and around the tail

plane, which affects the trim requirements. Thus, corrective control and trim is often required to maintain the desired rate of descent and airspeed.

A good landing is invariably the result of a well-executed approach, which in turn depends upon the maintenance of the desired approach slope at a constant angle. One method of achieving this is by using the perspective phenomenon. A runway appears to change its shape as the pilot's observation point changes. For example, seen from final approach a runway will appear wider at the approach end than at the opposite end. When a constant approach angle is maintained, the apparent configuration of a runway will also remain constant. The pilot sees the runway as a four-sided figure with the approach width much greater than that of the far end and the runway sides of equal length but converging toward the horizon. If the approach angle is made steeper, the runway will appear to grow longer and narrower. If the approach angle is made more shallow the runway appears to grow shorter and wider.

Although the runway area steadily grows larger as the approach progresses, as long as the relationship of the sides of the runway configuration remain the same the approach angle is remaining constant and touchdown will be near the threshold (Fig. 2-50).

Cross-Wind Landings

It is not always possible or practical to land directly into the wind. Consequently, the principles involved in cross-wind landings must be learned and practised until they present no difficulty. A significant change in wind direction is possible between the time an aircraft takes off and the time it lands, even during one circuit, so it is important that you are able to cope with crosswinds before your first solo.

An aircraft landing directly into the wind tends to maintain a straight heading while it is rolling on (or about to touch down on) the runway, with minimum control assistance from the pilot. However, in a crosswind, which is any wind affecting the aircraft at an angle to its longitudinal axis, a more complex situation exists, which if not properly attended to can cause a loss of control. The landing heading of an aircraft is normally determined by the direction in which the runway is oriented, rather than by the actual wind direction. Therefore, an aircraft landing in a crosswind has the wind striking it from one side or the other while it is in contact with the ground, and due to the aircraft's inherent tendency to weathercock it is being forced off its intended heading. Prior to landing, the aircraft will tend to drift across the runway instead of running true to the centre line. If no corrective

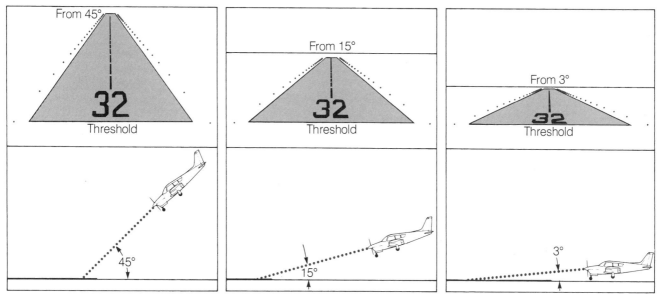

Figure 2-50   A Runway as It Appears from Different Angles

action is taken an undesirable side force is exerted on the landing-gear when it touches the surface. The same condition will occur if the path of the aircraft is held true to the centre line, with compensation for drift by crabbing, and the wheels allowed to touch the surface while not aligned with the direction of the runway.

Cross-wind landings are somewhat more difficult to manage than cross-wind take-offs. This is mainly due to the difference in the difficulties presented in maintaining control over the aircraft while speed is decreasing, instead of increasing as in the take-off. During take-off, as the speed of the aircraft increases, aerodynamic control of the aircraft becomes progressively more positive; as the aircraft's speed decreases, before and following touchdown, the effect of this control decreases. Before attempting a landing in a cross-wind, other than a very slight one, consult the Cross-Wind Component Chart (Fig. 2-38) in Exercise 16. It is used in the same manner as for a cross-wind take-off.

There are two basic methods for counteracting drift while executing a cross-wind landing.

The first, a side-slip, or wing down method of counteracting drift is probably the most popular of the two basic methods (Fig. 2-51). It affords the important advantage of continuity in flight control positioning, from before touchdown to the end of the landing roll, and will compensate adequately for acceptable cross-winds under most conditions. When using this method, avoid initiating the slip too far back on the final approach unless there are other reasons for slipping. As you approach the landing area and drift becomes apparent, side-slip into wind sufficiently to counteract the drift. Keep the longitudinal axis of the

aircraft aligned with the centre line of the runway by use of the rudder. On touchdown devote all possible attention to keeping the aircraft rolling in a straight line to forestall any tendency for the aircraft to ground loop. The aileron control should be held toward the upwind wing after touchdown to prevent it from rising.

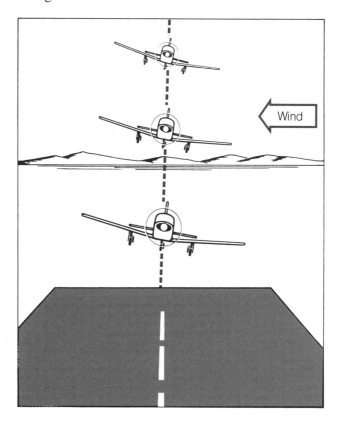

Figure 2-51   Landing in a Cross-Wind

When the side-slip method is used, the upwind main wheel will make contact with the surface first. The downwind wheel of the aircraft is still airborne but under normal circumstances the upwind wheel is not subjected to undue weight or impact stress.

The second method for eliminating drift when landing in a cross-wind requires considerable skill, excellent timing and a great deal of practice. For these reasons it is seldom used in elementary training. With this method the aircraft is maintained on a heading (crabbed) into wind so that the flight path of the aircraft is aligned with the runway centre line. This means that the longitudinal axis of the aircraft is not aligned with the intended landing path and if contact with the surface is allowed in this condition, there is a risk of damaging the landing-gear or subsequent difficulty in controlling the aircraft. Therefore, just prior to touchdown the longitudinal axis of the aircraft must be lined up with the runway by use of rudder. This method requires prompt and accurate rudder action to line up the aircraft exactly with its direction of travel over the ground at the instant of contact. If contact is made too soon the aircraft will land with crab; if contact is too late, it will land with drift. Either will impose side-loads on the landing-gear and impart ground looping tendencies. As well, as the upwind wing has not been lowered into the wind, a gust at the wrong moment can easily lift it and aggravate the tendency to groundloop.

Short Field Landing

Aircraft Flight Manuals usually describe the technique to be used for landing in a short field. They also provide tables or charts showing what landing performance you can expect. Compared to what you see in routine flying, this performance is impressive. If any head wind is present, you might expect a very short ground roll after landing, perhaps only a few hundred feet.

Achieving the performance figures given in the Aircraft Flight Manual requires careful handling of the aircraft and good judgement. The aim is to approach at the airspeed recommended for the aircraft weight and to touch down at the desired spot at the lowest possible airspeed commensurate with safety. Touching down at the required spot requires precise control of the approach so that you flare at the right point and at the right speed. The aircraft will normally float some distance after round-out, and this must be considered when aiming for a particular touchdown point.

Power

On a short field approach power is required to more accurately control descent. When power is used, very little change in the aircraft's pitch attitude is required to make necessary corrections in the approach slope. Leave power on until the landing flare is completed.

A high power, "drag in" approach is not recommended as a landing technique for a short field. This method can very easily bring the aircraft into slow flight and result in undershooting the landing area. It also requires that the pilot maintain skill in a second approach method when there is an obstacle. One method can serve perfectly well in all short field situations.

Touchdown

The aircraft should touch down on the main wheels at its lowest possible airspeed. Lower the nose wheel onto the runway as recommended in the Aircraft Flight Manual. Tail wheel types should be held in the three-point attitude for touchdown and braking.

Retraction of the flaps after touchdown is recommended by some flight manuals. While leaving flaps extended initially produces aerodynamic braking, it also produces lift that, by taking weight off the wheels, reduces the effectiveness of the wheel brakes. In aircraft with retractable landing gear, take care in raising the flaps during the landing roll to avoid unintentional gear retraction.

Obstacle

Whether a short field has an obstacle or not, most pilots prefer to use a power assisted approach. The aim is to establish a stabilized approach configuration and speed and maintain a constant descent angle to the flare (Fig. 2-52). The steepest angle of descent will be achieved with the throttle closed, but this can also give very high rates of descent that can be hard to check in the flare.

Wind Gusts

When approaching for a landing under strong wind conditions, the need for a reduced approach speed decreases as the wind speed increases. In gusty wind conditions it is advisable to add an amount equal to half the gust factor to the calculated approach speed. For example, if the wind is gusting from 15 to 25, the gust factor is 10; therefore, an approach that would normally be flown at 65 should be increased to 70 to

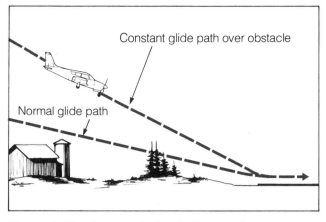

Figure 2-52 Approach Angle Comparison

allow for gusts. An airspeed increase of 10 is normally the maximum applied; increases above that could result in extremely long landing runs, and on some runways it may be impossible to stop before going off the end.

Decision making

Short field landings will challenge your decision-making skills. You have to decide whether the field length warrants a special technique. Most short fields are turf or gravel strips, so you may have to inspect the field first to decide whether landing is appropriate. You will have to consider wind direction, possible slope of the landing surface, obstacles, and the possibility of turbulence, wind shear, or rising or sinking air currents. If the field is really short, you will have to decide whether you will be able to take-off later, as it usually takes more distance to take off than it does to land. If the approach is not going well, you will have to decide early to overshoot and try again.

Soft and Unprepared Field Landings

Landing on a soft or unprepared field also requires a touchdown at the lowest possible speed consistent with safety. No special approach technique is needed. Assuming the length of the field is not a problem, a normal power-on approach works well, but care is needed on touchdown. Use of some power can help achieve a smooth landing. The attitude on touchdown should be nose up, roughly the attitude for a power-off stall. After touchdown, hold the nose wheel clear of the surface as long as possible and use brakes with care to prevent excessive loads on the nose gear. In tail wheel aircraft the tail wheel should touch down with or just before the main wheels and should be held down with the elevators throughout the landing roll.

The aircraft will decelerate quickly on a soft field. You might find it necessary to hold the control column fully back and add power as speed is reduced to keep the aircraft from getting stuck in the soft surface or nosing over.

Landing Run

The objective of any landing is to use as short a landing run as is consistent with safety and good judgement. Several factors that can lengthen the landing run must be taken into account. Identifying them requires knowledgeable observation. Then good judgement and skill are needed to compensate for them adequately. Some of these factors are:

Cross-Winds. The skills required for this factor have been adequately covered in this exercise; however, when contemplating a landing in a cross-wind, give due consideration to the diminished head-wind component. In fact, in cross-winds of 70 to 90 degrees other control factors make it necessary to make allowances in the length of the landing run as though a no wind condition existed.

Light, Shifting Winds. Under these conditions you may be wise to allow for a landing run length that the aircraft would experience under no wind conditions.

Tail Winds. Occasions may occur when you have no alternative to accepting a tail wind. Provided conditions permit, use the short field landing technique to reduce the ground speed at which the wheels make contact with the ground. To understand the effect of a tail wind on a typical light aircraft performing a normal landing on a hard surfaced runway with a pilot of average skill, consider the following.

With flaps extended until it comes to rest, and only moderate use of brake toward the end of the landing run, a certain aircraft touching down at 60 KT in a 20 KT head wind requires a landing run of 800 feet (244 metres). The same aircraft in a 20 KT tail wind requires a landing run of 1600 feet (488 metres), with moderate to heavy braking starting at the 800 foot (244 metre) point. When an aircraft must be landed in a tail wind, retract the flaps and place all controls in the neutral position as quickly as possible after the aircraft is positively on the ground.

Runway Gradient. When constructing an airport, the construction engineer does everything possible to provide a landing area without any gradient. However, this is almost impossible to do, and in many cases there may be a considerable downgrade or upgrade to a runway. This situation may be amplified considerably at remote airstrips or fields not primarily constructed as aircraft manoeuvring areas. Even a relatively imperceptible downgrade can increase the landing run considerably. When a downgrade is readily perceptible, all the factors must be assessed carefully and a decision to land uphill with a slight tail wind may have to be made. Some of the factors to consider are: degree of gradient, strength of wind, length of landing area, type and condition of surface, obstacles on approach, condition of aircraft braking system, and the skill and experience of the pilot.

Gross Weight.   Operating conditions being equal, the heavier an aircraft is, the longer a landing run it will require. The reasons for this fact are basic, but it is surprising how often it is ignored when a marginal landing area is being assessed. An unbraked light aircraft in still air [empty weight 1,000 pounds (454 kilograms), gross weight 1,500 pounds (680 kilograms)] will roll an additional 5.3 inches (135 millimetres) for every pound of load between its empty weight and its gross weight. With the pilot and half its fuel load [total fuel 70 gallons (265 litres)] it will roll, say, 500 feet (152 metres) after touching down. With a full fuel load and a passenger [260 more pounds (118 kilograms)], it will roll an additional 138 feet (42 metres). Under normal conditions this may not be too important, but with a short field, obstacles on the approach, a slippery surface, and a light and variable wind, it becomes increasingly significant.

Grass Surfaces.   A suitably sized, well-maintained grass area, free of soft and rough spots, probably offers one of the best landing surfaces for light aircraft. A grass surface helps absorb the shock of a hard landing and is much more tolerant of a landing made without sufficient compensation for cross-wind drift or imprudent sudden application of brake. In addition, even short clipped grass will offer significant resistance to the roll potential of the wheels, this resistance becoming greater as the aircraft slows down. The end result is that a grassy surface which is suitable in other respects, will afford a much shorter landing roll than you could expect from a paved surface. However, if you are making a landing into a short grassy field and you know that braking action will be required, it is also important to know that if the grass is wet, or even damp with dew, the wheels will skid very easily and provide up to 30 per cent less braking capability than a wet paved surface. One rule of thumb is: the greener the wet grass the more slippery the surface. To illustrate this point, float planes have landed and successfully taken off again on wet, lush grass.

Wet Runways **Hydroplaning.**   When hydroplaning occurs, the tires of the aircraft are completely separated from the runway surface by a thin film of water. They will continue to hydroplane until a reduction in speed permits the tires to regain contact with the runway. This speed will be considerably below the speed at which hydroplaning commences. Under these conditions, tire traction drops to almost negligible values, and in some cases the wheel will stop rotating entirely. The tires will provide no braking capability and will not contribute to the directional control of the aircraft. The resultant increase in stopping distance is impossible to predict accurately, but it has been estimated to increase as much as 700 percent. Further, it is known that a 10 KT cross-wind will drift an aircraft off the side of a 200 foot (61 metre) wide runway in approximately 7 seconds under hydroplaning conditions. When hydroplaning is suspected, release the brakes immediately, then reapply a very slight pressure. This pressure may be gradually increased as the aircraft slows down, but be prepared to release the brakes and reapply pressure as often as necessary.

Wheelbarrowing.   On landing, take care not to allow the nose wheel of a nose wheel equipped aircraft to touch the ground first, or simultaneously with the main wheel, if compensating for drift in a cross-wind. The nose wheel is not structured to bear the landing impact load nor accept major surface irregularities it may encounter at the high touchdown speed. If the nose wheel is steerable it may be cocked to one side when compensating for drift and if it is allowed to contact the runway at touchdown speed in this position, the aircraft may develop a swing and/or the nose wheel may be damaged. Nose wheel type aircraft are normally landed using a procedure in which the nose wheel is held off the ground as long as practical unless any difficulty is experienced in maintaining direction. Nose wheel type aircraft should not be flown onto the runway with excess speed, but under gusty and turbulent conditions the nose wheel may be lowered to the ground sooner than usual to prevent the aircraft from skipping or lifting off again.

You may remember that wheelbarrowing is a condition encountered when the main wheels are lightly loaded or clear of the runway and the nose wheel is firmly in contact with the runway. This causes the nose gear to support a greater than normal percentage of weight while providing the only means of steering. In the extreme condition, a loss of directional control may ensue at a very critical point in the landing procedure. Wheelbarrowing may occur if the aircraft is allowed to touch down with little or no rotation and the pilot tries to hold the aircraft on the ground with forward pressure on the elevator control. This often occurs as a result of the use of excessive approach speeds, particularly in a full flap configuration. In a cross-wind the aircraft in this situation tends to pivot (yaw) rapidly about the nose wheel, in a manoeuvre very similar to a ground loop in a tail wheel type aircraft. Other indications of wheelbarrowing are wheel skipping and/or extreme loss of braking effect when the brakes are applied.

Wheelbarrowing incidents have occurred during cross-wind landings in aircraft equipped with nose wheel steering when the "slip" technique for cross-wind correction is being used. On many training aircraft the nose wheel steers when rudder is applied; for this reason, such landings require careful rudder operation just prior to and during touchdown.

Corrective Action (Wheelbarrowing). Corrective action must be based on a number of factors, i.e., degree of development of the condition, the pilot's proficiency, remaining runway length, and aircraft performance. After considering these factors, you should initiate the following corrective measures:

1. Close the throttle, relax forward elevator pressure to aft of the neutral position to lighten the load on the nose gear, and return steering and braking to normal. If the flaps can be retracted safely, additional braking will be obtained on dry runways.
2. If control can be regained and adequate aircraft performance and runway are available, abort the landing and go around again.

A *ground loop*, generally associated with tail wheel equipped aircraft, is defined as: "A violent uncontrollable turn resulting from failure to correct (or overcorrecting) a swing on landing (or take-off)." An undesirable turn during the ground operation of an aircraft is generally referred to as a *swing*. A swing may be caused by any of the following:

1. Touching down while crabbing into wind.
2. Touching down when the aircraft is drifting sideways.
3. A cross-wind acting on the fuselage and rudder, causing an aircraft to weathercock into wind.
4. Allowing the upwind wing to rise, which combined with weathercocking effects on the tail causes a swing into wind.
5. Failing to control a wheel landing properly: when the tail settles onto the runway, heading control changes from rudder to tail wheel and during the transition period, a ground swing can develop.
6. Incorrect recovery action for drift after a bounce, which has the same effect as landing with drift.

To prevent a swing from developing into a ground loop in a tail wheel aircraft, you must anticipate and take firm, immediate action. Keep the control column fully back and apply opposite rudder.

If the landing is doubtful, or if you are starting to get into trouble, open the throttle and go around again, as long as you are positive there is adequate runway or open space to do so. In gusty cross-wind conditions, retract the flaps as soon as the aircraft is firmly on the ground. Flaps provide more surface for a cross-wind to act upon; therefore, if they are retracted there will be less swing effect.

Air Density

The Aircraft Flight Manual usually contains tables indicating the effect that airport elevation and ambient temperature have on the length of the landing run. *Density altitude* is the key term in determining the length of landing run required; it may be reviewed by referring to Exercise 16.

The required landing run length of an aircraft is based on its performance in standard atmosphere — i.e., 29.92 inches of mercury and an ambient temperature of 15°C. Rules of thumb of the effect of temperature and pressure varying from standard are:

1. If the air is warmer the landing run will be longer.
2. If the aerodrome is higher the landing run will be longer.
3. If the air is warmer and the aerodrome is higher the landing run will be even longer.

Wheel Landings

As you gain experience, you may find that the ability to carry out wheel landings in tail wheel type aircraft will be beneficial to you when flying specific types or under certain landing conditions. The approach should be normal with or without power according to the conditions of the day, to the point where the descent is checked. The airspeed is then decreased to the point at which the aircraft settles. Adjust the power at this point so as to descend in a level attitude at a slow rate (approximately 100 to 300 feet per minute). You will not be able to watch the vertical speed indicator during this stage, but with practice you can easily estimate the descent rate. A fast rate of descent could cause a hard contact with the surface, followed by a downward rotation of the tail through inertia and a subsequent bounce back up into the air. Small control adjustments only should be used as the aircraft settles to assist in descending slowly and maintain a level attitude. As the wheels smoothly contact the surface, apply gentle but firm forward pressure to hold the wheels on the ground and decrease the angle of attack. The aircraft should be held on the wheels, nearly level, until it has slowed sufficiently to ensure full control in a three-point attitude under existing conditions.

As speed gradually decreases in the landing roll, a transition point is reached in tail wheel aircraft at which the rudder ceases to provide adequate directional control. At this point the tail wheel must be positively and firmly on the ground so that the aircraft may be directionally controlled by the action of the steerable tail wheel. Keep the control column well back until the aircraft is clear of the runway and in the taxiing mode.

Recovering from Bad Landings

It doesn't take very much to turn a promising landing into a bad one. The sequence of events often starts with a poor approach, then all it takes is a gust of wind, or a tendency to overcontrol the aircraft, or touching down while the speed is too high. Even combinations of these are possible, and quick recognition of the situation and careful handling is needed to recover.

You first have to recognize that the situation calls for corrective action. Then you have two choices. You can either go around or you can continue the landing. Your choice will be dictated to some extent by the airspeed, the height of the aircraft, and by your own skill.

An aircraft can touch down and immediately return to flight a mere two or three feet (1 metre) above the runway, or you could find yourself at ten to twenty feet (3 to 6 metres). It can "balloon," which means the aircraft rises due to overcontrolling or to a gust of wind in the flare. You might also round-out too high. All these situations result in an aircraft that is flying close to the ground with airspeed that is very low and getting lower.

If you see a bad landing developing, hold the elevator controls steady. Do not move them for a moment until you see what is going to happen. Whatever you do, *do not* immediately apply forward pressure. This will only result in an immediate loss of lift, a rapid rate of sink, and hard contact with the ground, probably in an attitude that could damage the nose wheel. If you are flying a tail wheel aircraft this will likely result in a series of bounces caused by overcontrolling with each successive landing becoming worse than the one before.

After you check the elevator controls and if you judge that the rise in altitude is not serious, perhaps 5 feet (2 metres) or less, then wait for the aircraft to settle before continuing with the landing. Remember that the airspeed will continue to decrease and when you are ready to continue holding off, larger control movements will be needed to stop the descent.

However, if you sense you are somewhat higher then apply power, maintain control and slowly ease the nose down in the same manner used when recovering from slow flight. Do not overcontrol. Make minor adjustments as necessary to stop the sink. Even if the aircraft lands hard, provided that it lands on the main wheels, it is unlikely that any damage will occur.

While you are taking corrective action you must remember to continue to eliminate any drift.

If you aren't sure that you can continue the landing safely, you should go around. As you may be in the slow flight speed range, going around will have to be done carefully. You will need immediate yet smooth application of full power and control of yaw. Carburettor heat will be set to cold and flaps will be brought up in stages according to the procedure in the Aircraft Flight Manual.

Overshooting (Going Around)

The decision to overshoot or go around because of a poor approach or landing rests with the pilot. Occasionally, however, the control tower may ask you to go around. The tower will normally issue this instruction by saying "Pull up and go around."

As soon as the decision to overshoot has been taken, apply full power, accelerate to a safe climb speed in level flight, reduce flap extension as required according to type, and raise the nose to the climbing attitude. Keep straight as the throttle is opened, and roughly trim off the pressure on the control column. Start the climb, and when you have firm control of the aircraft, raise the flaps, adjust the climbing speed, and retrim.

If a bad approach, flare, or landing is the cause of the overshoot, and the remaining portion of the runway is clear of other aircraft, it is permissible to climb straight ahead. However, if you have been forced to go around for some reason while on the approach, it is difficult to see ahead and below. Ease over to the right of the runway, fly parallel to the runway, and while climbing out keep a look-out for other traffic, especially aircraft taking off and climbing up beside you.

It is a general rule that if carburettor heat is in the "on" position during the approach to landing, it should be placed in the "off" position as soon as possible after power is applied on the overshoot procedure. However, be guided in this matter by the Aircraft Flight Manual and/or the rules of the Flight Training Unit.

Wake Turbulence

Wake turbulence, generated by preceding larger aircraft, should be avoided by light aircraft at all times but especially during approaches and landings.

Since vortices are subject to many variable factors (size, weight, and speed of the aircraft and air conditions) it is not possible to forecast their presence accurately. However, it should be remembered that the vortices are carried by the ambient wind and have a downward movement imparted to them when they are shed, and an outward movement near the ground, due to cushion effect.

1. When it is necessary to operate behind a large heavy aircraft, remain above the flight path of that aircraft.
2. When preparing to land remember that wake turbulence generated by a preceding aircraft will be maximum:
   (a) just before the point of touchdown for a landing aircraft; and
   (b) at the point of lift-off for a departing aircraft.
3. On landing, wake turbulence can best be avoided:
   (a) if following an aircraft that just departed, by planning your approach to land near the approach end of the runway so as to be down before reaching the point where the preceding aircraft took off; and
   (b) if following an aircraft that has just landed, by planning your approach so as to stay above the flight path of the preceding aircraft and to touch down beyond the point where the preceding aircraft touched down (Fig. 2-53).
4. Remember, even though a clearance for take-off or landing has been issued, if you believe it is safer to wait, to use a different runway, or in some other way to alter your intended operation, ask the controller for a revised clearance.

Visual Illusions

Visual illusions are frequently mentioned by accident investigators as contributing factors in approach and landing accidents. How do these illusions occur? Research has thrown light on these phenomena.

In approaching a runway, you will normally use several cues to adjust your glide path. One important cue is the visual angle between the point you are aiming for on the runway and the horizon. This angle is equal to the angle of approach.

If the horizon is not visible, because of poor visibility or darkness, a number of other means may be used to estimate your position. Your perspective of the shape of runway is such a cue (Fig. 2-50). The extended sides of the runway intersect at the horizon, and the surface texture and objects of the surrounding

terrain also suggest the position of the horizon. But these cues are only accurate when the terrain and runway are level. Sloping terrain and runway can give you a false sense of the horizon and mislead you into flying an approach that may be too steep or too shallow for safety.

Even a relatively small upslope of the runway can give the illusion that you are higher than you should be and mislead you into making a shallower than normal approach. This may place you dangerously close to objects on the approach and may cause you to touch down before the runway threshold (Fig. 2-54).

Similarly, when the runway slopes away from your landing aircraft, you may fly a steeper than normal approach. Sloping terrain on approach to the runway can have the same deceptive effect as a sloping runway. It would not be unusual for the approach terrain and the runway to slope in the same direction and so these two different effects would then be cumulative.

If an approach is made over water, snow, or other featureless terrain, or carried out over a darkened area, there is a tendency to fly a lower than normal approach. Factors that obscure visibility such as rain, haze, or dark runway environment can have the same effect.

Bright runway and approach lights may cause you to believe that the airport is closer than it actually is. The result may be a premature descent, but this depends on how you perceive this situation and what other illusions are present.

Steep surrounding terrain can create an illusion of being too low and can lead to a steep approach.

The round-out and touchdown phase of the landing require you to make important height judgements. To do this you are likely to use a number of cues. These include apparent speed, apparent ground texture, and size of known objects. If these cues are removed because of an approach over water, snow, featureless terrain, or darkness, you may be forced to make a judgement based on only one cue, for example, runway width.

If you are used to a runway 100 feet (30 metres) wide and approach a narrower runway, say 50 feet (15 metres), the smaller target may give the illusion that you are higher than you should be, influencing you to descend to a lower than normal altitude before round-out. If the error is not corrected in time, this could result in a hard landing.

Conversely, if the unfamiliar runway is significantly wider, say 150 feet (45 metres), you may experience the illusion of being lower than you actually are. This would influence you to make a higher than normal round-out, and could also result in a hard landing.

To take preventative measures against visual illusions it is important to realize that visual illusions are normal phenomena that affect all of us. Illusions result

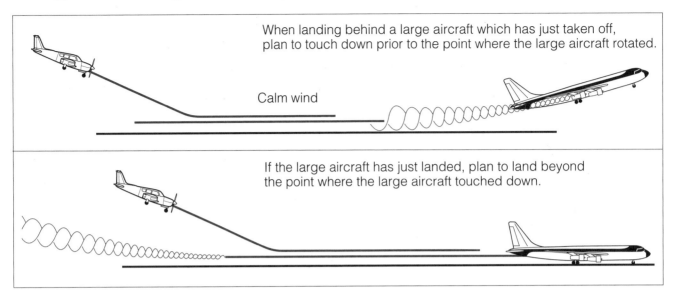

When landing behind a large aircraft which has just taken off, plan to touch down prior to the point where the large aircraft rotated.

Calm wind

If the large aircraft has just landed, plan to land beyond the point where the large aircraft touched down.

Figure 2-53   Landing behind a Large Aircraft

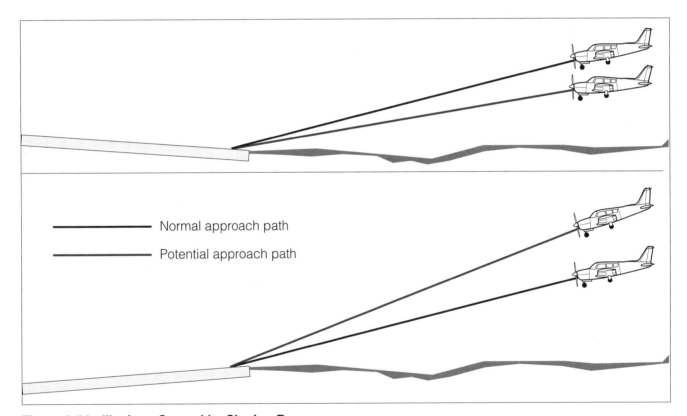

Normal approach path

Potential approach path

Figure 2-54   Illusions Caused by Sloping Runways

from the way humans process visual and other information. Therefore, you must understand the nature of the illusions and the situations in which they are likely to be encountered.

When you expect an illusion, an effective counter measure is to supplement visual cues with information from other sources. In making visual approaches to unfamiliar aerodromes, note the size and terrain conditions of those destination aerodromes when planning

the flight. Plan to take advantage of visual aids such as VASI lights (where available). It is often helpful to overfly an unfamiliar field before making an approach to land, and ensure that you fly a normal circuit. Often things such as the slope of a runway cannot be accurately judged from the air, but your awareness of situations where you expect illusions and knowledge of their effects will go a long way in helping you to make safe decisions to deal with them.

Touch-and-Go

To save flight time *touch-and-go* landings are frequently employed during various stages of training. In this manoeuvre a take-off is executed while the aircraft is still in its landing roll. This means that the essential components of a pre-take-off cockpit check must be carried out while the aircraft is moving along the runway. Before attempting the cockpit check be sure that the aircraft is under complete control in the landed mode. The essential components of touch-and-go pre-takeoff actions are:

1. Flaps up (or set for take-off).
2. Trim set for take-off.
3. Carburettor heat "cold."
4. Power "full" (or take-off setting).

When operating at a controlled airport a standard landing clearance presumes that you will land and exit from the runway. If a touch-and-go landing is intended you must obtain clearance to do so from the control tower. Their clearance will most likely be "cleared for touch-and-go."

Clearing the Runway

Unless otherwise instructed by ATC, aircraft are expected to continue in the landing direction to the nearest suitable taxiway and exit the runway without delay. No aircraft should exit a runway onto another runway unless instructed or authorized to do so by ATC. When required, ATC will provide the pilot with instructions for leaving the runway. These instructions will normally be given to the pilot prior to landing or during the landing roll. After landing on a dead-end runway, a pilot will normally be given instructions to "back track." After leaving the runway, unless otherwise instructed by ATC, pilots should continue forward across the taxi holding position lines or to a point at least 200 feet (61 metres) from the edge of the runway where a taxi holding position line is not available. The aircraft is not considered clear of the runway until all parts of the aircraft are past the taxi holding position line or the 200 foot (61 metre) point.

Simultaneous Intersecting Runway Operations (SIRO)

To increase the traffic capacity of an airport, ATC may use special procedures for simultaneous use of intersecting runways. For example, if an aircraft is landing while another aircraft is either approaching to land or taking off from an intersecting runway, the clearance to land will include an instruction to "hold short" of the intersecting runway. If the pilot is unable to comply with this instruction, the clearance must be declined. Knowledge of your aircraft's performance and your personal skill level is very important when determining your ability to comply with these instructions. There are a number of conditions under which Simultaneous Intersecting Runway Operations (SIRO) may be approved. These can be found in the RAC section of the *Aeronautical Information Publication* (A.I.P.) Canada.

First Solo

The first solo is a landmark in your flying career. You will never forget it, and it is quite normal to look forward to it, but do not exaggerate its importance. It is not so much when you solo, but rather what you know and what you can do correctly at this period of training that is important. Soloing is merely another step in the orderly process of flight training, bringing you to the stage where learning really begins.

The amount of dual instruction required to solo need not be a reflection on your ability. Everyone varies in capacity to learn, and very often the student who is a little slow to learn ultimately makes the better pilot.

Before being permitted to take your first solo flight you will have to satisfy your flight instructor that you are able to:

1. Take off and land while using the correct technique for runway surface and wind conditions.
2. Fly accurate circuits while maintaining safe separation from other aircraft.
3. Correct a potentially poor landing, but be capable of judging when it is necessary to go around again.
4. Recognize whether you are overshooting or undershooting a predetermined touchdown zone and take early corrective action.
5. Operate the radio competently where airport traffic control is in effect, and in the event of communications failure know the emergency procedures to follow and the light signals that may be directed to you from a control tower.
6. Use the correct overshoot procedures.
7. Realize the importance of keeping an alert lookout for other aircraft now that you are the sole occupant of the aircraft.
8. Conduct a forced landing from any point in the circuit in the event of an engine failure.
9. Handle the emergency procedures listed in the Aircraft Flight Manual.

10. Properly adjust your circuit pattern in the event of a change of runway in use after you take off.

As pilot-in-command you are responsible for the operation and safety of the aircraft when on solo flight. However, for this first solo your instructor will ensure that suitable conditions exist and precautions are taken by:

1. Ensuring that the aircraft has sufficient fuel for the intended solo flight, with adequate reserve for possible delays or overshoots.
2. Allowing the flight only when sufficient daylight must remain for successful completion of the anticipated flight, allowing a liberal margin for possible additional circuits due to traffic congestion, overshooting, etc.
3. Ensuring that the first solo is flown only if suitable weather conditions exist and are forecast to continue.
4. Selecting a time when air traffic conditions are light.
5. Ensuring that the solo flight does not follow a lengthy session of dual instruction.

Securely fasten the seat-belt in the empty seat.

On a solo flight, the take-off will be much quicker and the climb more rapid due to the absence of the instructor's weight. Many students have remarked that this was the outstanding feature of their first solo flight: they were not fully prepared for the suddenness with which the aircraft became airborne.

Since the aircraft is relatively lightly loaded, it will require less power to maintain a desired rate of descent. Also, after the flare for landing, the aircraft will tend to "float" longer before touching down, and it may be more sensitive to gusts during the initial stage of the landing roll.

New manoeuvres and procedures will be added as progress permits and further solo periods planned and authorized. Specific practice on procedures learned in earlier stages of your training will also be included. When authorized solo to do specific manoeuvres, it is important to practise the specific work diligently; there will be ample time for sightseeing or pleasure cruising when flight training is finished. Perfection of technique as early as possible is your objective; therefore, after the first solo, subsequent solo flights must be devoted to attaining greater precision, co-ordination, orientation, and judgement.

Illusions Created by Drift

There are times when it may be necessary to manoeuvre an aircraft relatively close to the ground, such as during a forced landing, when carrying out precautionary landing procedures, or because of deteriorating weather. On these occasions it is very important that you recognize and understand illusions created by drift (Fig. 2-55).

Once an aircraft becomes airborne, it enters a medium of movement almost unrelated to any it encounters on the ground. As soon as the wheels leave the surface there is no wind in the sense that we normally associate with the word. Instead, the aircraft enters a body of air, and while airborne its movement is directly related to the speed and direction of movement of that body of air.

In flight at normal operating altitudes the movement of the aircraft relative to the ground appears to be comparatively slow even when the airspeed is quite high. However, when the aircraft is flown closer to the ground, movement in relation to the ground becomes more apparent and in strong winds illusions are created. If misinterpreted, they can develop into potentially dangerous flight conditions.

In conditions of good visibility, flight below normal operating altitudes can be at normal cruising speed, but if it is necessary to fly near the ground in reduced visibility it is usually advisable to reduce speed. The flaps should be partially extended when flying at lower speeds near the ground. This will allow a lower operational speed, a smaller turning radius to avoid obstacles, and a better view, owing to the lower position of the nose. The increased power required with flap extension will also improve control, due to the additional slipstream over the elevators and rudder.

To demonstrate illusions created by drift, your instructor will choose a day when the wind is strong enough for the effects to be easily discerned. Flying upwind the reduction in ground speed is noticeable. Flying downwind the increased ground speed is very noticeable, sometimes to the extent that there is a temptation to reduce airspeed, which if carried to extremes could lead to a stalled condition. Flying cross-wind, the sideways drift over the ground is very apparent, especially when the aircraft is aligned with a straight road or section line.

In a turn from upwind to downwind, because of the drift over the ground the aircraft seems to be slipping inward, even though the turn is accurate and well co-ordinated. This impression is an illusion and you must not use the rudder in the attempt to correct it. A quick glance at the centred ball of the turn-and-bank indicator will confirm that no slip is occurring. However, the drift itself is very real and plenty of room must be allowed when turning from upwind to downwind if there is an obstacle on the inside of the turn. Similarly, in a turn from downwind to upwind the aircraft seems to skid outward, although the ball is centred. This too is an illusion, but again the drift is real and ample room must be allowed between the aircraft and obstructions on the outside of the turn.

Remember, the lower the airspeed, the greater the illusion of skid and slip with a given wind velocity.

Low Flying.   The closer you are flying to the ground, the greater the illusory effect. At such times, little attention can be given to the flight instruments. Therefore, it is most important to understand the false impressions that can be created by the deceptive appearance of the ground. An alert watch must be maintained not only for other aircraft but also for high obstacles on the ground, which under the circumstances add considerably to the hazard. Continual vigilance is essential.

Since the aircraft's altimeter indicates the height of the aircraft above sea level and not its height above the ground, it is most important to watch the ground contours carefully and learn to estimate the height above the ground. Heights can be more easily judged by looking well ahead.

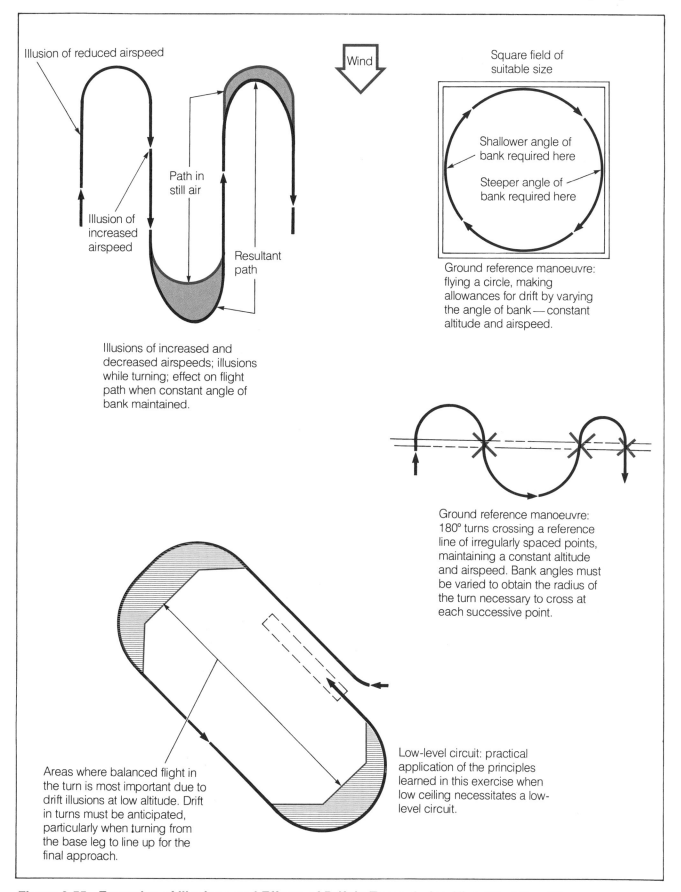

Illusion of reduced airspeed

Illusion of increased airspeed

Path in still air

Resultant path

Illusions of increased and decreased airspeeds; illusions while turning; effect on flight path when constant angle of bank maintained.

Wind

Square field of suitable size

Shallower angle of bank required here

Steeper angle of bank required here

Ground reference manoeuvre: flying a circle, making allowances for drift by varying the angle of bank—constant altitude and airspeed.

Ground reference manoeuvre: 180° turns crossing a reference line of irregularly spaced points, maintaining a constant altitude and airspeed. Bank angles must be varied to obtain the radius of the turn necessary to cross at each successive point.

Areas where balanced flight in the turn is most important due to drift illusions at low altitude. Drift in turns must be anticipated, particularly when turning from the base leg to line up for the final approach.

Low-level circuit: practical application of the principles learned in this exercise when low ceiling necessitates a low-level circuit.

Figure 2-55   Examples of Illusions and Effects of Drift in Turns during High Wind Conditions

Map Reading.   When an aircraft is flown at lower than normal altitudes, map reading becomes more difficult due to the reduced area of ground visible and the shorter time available for identifying landmarks.

Points to Remember

When flying close to the ground:

1. Maintain a safe airspeed.

2. Turn accurately in spite of the illusory effect of drift.
3. Maintain a safe height above ground contours.
4. Keep a good look-out.
5. Do not turn too steeply.
6. Do not annoy others or frighten livestock.

Never practise this exercise unless there is a flight instructor on board the aircraft. The height and suitability of the area should be governed by local restrictions and *Canadian Aviation Regulations*.

Precautionary Landing

The aim of this exercise is to successfully carry out a landing at a location where there is doubt as to the suitability of the landing surface or where advance information is unavailable. Therefore, the area and surface must be visually inspected to provide sufficient information for the pilot to make the best decision on circuit, approach, and landing procedures.

Each year, many landings are completed at sites that do not offer advisory service or published information. Pilots planning flights to these locations must find alternate sources of information to make the decisions associated with safe and efficient landing operations. In some situations, this information is available only from aerial observations.

Precautionary landings may result from a planned landing at a location about which information is limited, from unanticipated changes during the flight, or from abnormal or even emergency situations. When the procedure results from abnormal or emergency situations, the pilot is well advised to begin the precautionary landing early. The sooner the pilot locates and inspects a potential landing site, the less the chance of additional limitations being imposed by worsening aircraft conditions, deteriorating weather, or other factors.

Once the decision has been made to complete a precautionary landing, the pilot should consider completing associated radio calls and cockpit checks while at altitude, in order not to interfere with the inspection of the landing site.

Although suggested procedures are provided later in this chapter, pilots must understand that no two precautionary landing situations are identical. Wind may be different. Cloud conditions may vary. Visibility may change. In the case of an emergency situation, the urgency of the landing adds its own pressure. To complete a successful precautionary landing, the pilot must combine the skill to control the aircraft with knowledge of procedures and terrain.

A precautionary landing is an excellent opportunity to practise new and previously acquired skills. Before this exercise can be completed meaningfully, the pilot must possess basic flight skills associated with circuit procedures and all types of landings.

In its simplest form a precautionary landing involves two parts (Fig. 2-56):

1. A normal circuit flown to a low approach over the intended landing site to visually inspect the potential landing area.
2. Another normal circuit ending in a safe landing.

While inspecting the landing site, the following factors must be considered and evaluated.

Wind Velocity

Smoke gives a good indication of wind velocity. Dust is blown by the wind. Tall grass and crops ripple in the direction of the wind. The upwind side of bodies of water usually have a calmer surface, and wind lines are obvious in strong wind conditions. Some deciduous trees often display the white or silver coloured underside of leaves on the upwind side.

In the absence of any indication of wind, you will have to make the best possible estimate. To do this, remember the wind conditions at the point of departure and the forecast for your destination. Consider the ground speed and in-flight drift to give an indication of the wind aloft. Then compare the wind aloft with what would be expected on the ground.

Landing Area

To safely land an aeroplane, you need a surface that is:

1. Sufficiently long.

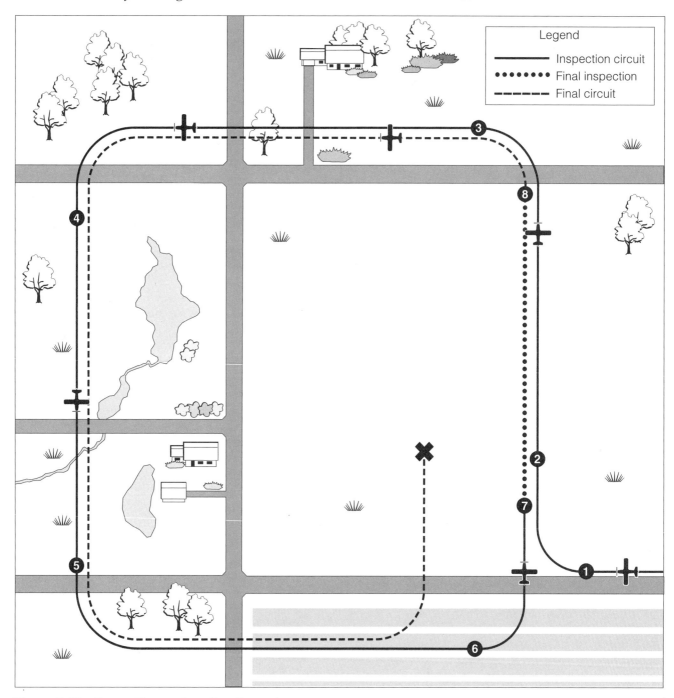

Figure 2-56 Precautionary Landing Procedure

2. Smooth and firm.
3. As level as possible.
4. Free of obstacles.
5. Into wind, if possible.

Transportation, communication facilities, and the need for assistance may be a consideration in the choice of landing areas.

You probably will be interested in departing later, so if possible, you will need a field that is adequate for take-off. Remember, it is not uncommon to require greater distances to take off than to land. As well, remember that changing conditions may turn a suitable field into an unsuitable one.

Except in emergency situations, pilots are not at liberty to use an area indiscriminately for landings or take-offs. When contemplating a landing at a private aerodrome, remember that "private" means private property and that permission from the owner or operator may be required before using the aerodrome. If you are proposing to land away from an aerodrome, remember that someone owns the land. The owner may not be enthusiastic about aircraft using the property. *Canadian Aviation Regulations*, provincial laws, municipal bylaws, and the rights of a property owner may limit a pilot's selection of a landing site.

Detecting Obstacles

Obstacles such as wires and towers are difficult to see, especially in conditions of poor visibility. Many obstacles appear camouflaged against ground terrain. These obstacles may not be readily seen until descent below their altitude makes them more visible against the sky. Take care to observe the sorts of things that could present problems in the circuit area, on final approach, and in the overshoot and departure areas.

Experience and observations up to this point will help you to detect obstacles. For example, you have probably noticed that power or telephone lines parallel many roadways; trees frequently border fields and laneways. As you fly or while driving through the countryside, pay attention to the location and type of potential obstacles.

Many hazardous obstacles look considerably different from the air than they do from the ground. Through the course of training your instructor will show you what obstacles such as trees or power lines look like from the air.

It may be useful to walk or drive along an area that you have inspected from the air. This ground level observation may help to better visualize height, distance, and surface conditions.

Airspeed

The speed chosen for the inspection of the landing area is the result of a trade-off. The slower you go, the more time you will have to inspect the landing area, but the more difficulty you will have controlling the aircraft. The faster you go, the more ease you will have in controlling the aircraft, but the more difficult it will be to see the surface detail. Every aeroplane has a minimum safe operating speed. In training thus far you have explored flight characteristics at and near the stalling speed. The aircraft is controllable in slow flight, but takes much more attention to operate in this range. Therefore, slow flight is not a good speed range to choose. As the primary objective is to inspect the landing surface, you must fly the aircraft in a way that requires a minimum of attention to maintain control leaving maximum attention to inspect the surface.

Most manufacturers recommend a speed in the lower portion of the normal operating range but above endurance to keep out of the slow flight speed range.

They may also suggest the use of partial flaps to give a more nose-down attitude than that normally associated with flight at slower speeds and to reduce the stall speed. In the absence of a speed given by the manufacturer, consider using a speed close to the normal approach speed.

Altitude

Altitude also presents some trade-offs. The lower you fly, the closer you are to what you are inspecting, and the greater the chance of identifying details on the surface. However, remember that as an aircraft gets closer to the ground, the sensation of speed increases (see Exercise 20). At the same time, the lower you go, the greater the likelihood of getting too close to obstructions such as trees, power lines, towers, or hills. The higher you fly, there is less chance of encountering obstructions, but your ability to see detail on the surface decreases. A compromise is required. If the procedure is being completed at an airport, a lower inspection altitude might be selected than would be used in a hilly, forested area. The flight training unit where you are training may have minimum altitude restrictions for practising these types of exercises.

Whatever altitude you select, take care to ensure that the aircraft is established in level flight at that altitude. Trimming the aircraft for flight at the selected speed is one way to reduce cockpit workload while maintaining aircraft control.

Inspection of the Landing Surface

If you fly directly over the landing area, you will be unable to see below. You must fly to one side. If flying an aircraft with tandem seating, fly along whichever side of the landing area you like. If in the left seat of a side-by-side seating aircraft, flying along the right side of the landing area will allow the best view of the landing area.

Fly close enough to the proposed landing path to see it plainly and far enough to the side to have a clear view without having to look through wheels, struts, or other parts of the aircraft. If flying a high wing aircraft, there are fewer aircraft parts to interfere with downward vision, and you may be able to fly closer to the landing area than in a low wing aircraft. The best place to fly is along a path that gives the best view of the potential landing surface.

The inspection work associated with precautionary landings can be completed at low or high altitudes. If appropriate, both high and low passes provide their own important information.

Some pilots suggest that when both a high- and low-level pass are planned, the high pass should tell you about the particular flight path to follow. It should also let you identify any obvious reasons not to land at the location. Subsequent low-level passes must provide good reasons to land at the site.

Sometimes conditions require a number of inspection passes. There may be much to see, many potential problems with the chosen landing area, and little or no advance information about the landing site. In situations where advance information about the landing site is available, fewer or even abbreviated inspection procedures may be appropriate.

When inspecting an unknown area in which there are numerous obvious obstacles, some pilots suggest a number of inspection passes at progressively lower altitudes to prevent encountering unexpected obstacles on an initial low pass.

During the inspection pass, speed and altitude should be stabilized so that the aircraft can be controlled with as little of the pilot's attention as possible. The pilot can then devote maximum attention to inspecting the field. Many pilots at some point in their training determine the power setting required for level flight at the inspection pass airspeed. When reaching the area to be inspected, the pilot merely has to set the known power and trim for level flight at the desired airspeed.

Your instructor will demonstrate altitudes, speeds and flight paths that are appropriate for the inspection.

Length of the Landing Surface

There are methods you can use to determine the length of the landing surface. In some areas, roads are known distances apart and fields are of known length. For example, much of Western Canada features roadways that are either one or two miles apart and fields that are generally one-half mile or one mile square. If operating in these areas, you have a distance scale to determine the length of the potential landing site. You may now compare ground distances in miles [approximately 5,000 feet (1500 metres)] with figures obtained from aircraft landing performance data.

If these natural scales are not available, compare the potential landing path with runways on which you have landed. Exercise care when a comparison to a known runway is the only basis for determining length because of factors such as optical illusions and the indirect comparison.

Your speed during the inspection pass can provide assistance. An aircraft travelling at a ground speed of 60 KT covers one mile in one minute or approximately 100 feet each second. In this case, to calculate field length check the time in seconds and multiply by 100. This provides another means of judging field length provided your calculations are based on a reasonably accurate ground speed estimate.

Use as many means of determining length as possible. If the lengths from each system are similar, you are justified in believing the figure. If there is significant discrepancy, additional caution is warranted.

Suitability of the Landing Surface

You must consider whether the field is smooth, level and hard enough to support the aircraft. Aside from the surface itself, you must determine whether any vegetation growing on the surface will create a hazard during landing and subsequent take-off. Some crops are thick, tall or intertwined enough to cause considerable resistance to the forward motion of an aircraft.

You must rely on your knowledge of what acceptable surfaces look like. This is the time to look for features that experience tells you can indicate problems. For example, a dark coloured field with signs of having been recently cultivated is probably quite soft. Standing pools of water, snow-drifts and evidence of deep snow are also signs of a soft landing area. Rocks, holes, or furrows in a field may indicate a surface too rough for use by an aircraft. Shaded areas or areas that appear to rise above or fall below the remainder of the surface may well indicate holes or mounds that can cause problems. Patches of different coloured vegetation can indicate hollows, wet areas, or even overgrown obstructions.

Knowledge of what has been happening recently in the area will help you in selecting a landing site. After significant rain, areas at the base of hills are more likely to be moist and soft than areas at the top. During certain times of the year, agricultural fields may have been cultivated and are probably quite soft. At other times, heavy harvesting equipment may have been on the field indicating surfaces hard enough to support an aircraft.

During training, your instructor will point out features that indicate suitable and unsuitable surfaces.

Touchdown Area

If your chosen landing site is relatively short and narrow, there is little choice as to where you touch down. However, if your landing area is longer than your aircraft requires and the initial section of the runway is obviously too soft or there are obstacles, you may decide to touch down at some point farther along.

If you are landing in a large square field that appears to be acceptable throughout, you will need to decide on a landing path. Perhaps a diagonal landing will allow you to land into wind. Maybe landing parallel to the edge of the field near habitation and accepting a slight cross-wind is better.

Approach and Landing

Your training to this point has included procedures for landing on a variety of fields. This exercise involves the practical application of that training. The information from your inspection will indicate whether or not the selected location is suitable for a landing.

Once you have decided to land at a particular location and identified the precise landing path, you need to decide the best approach procedure and aircraft configuration. You will have to consider wind, surface, obstacles and touchdown point to determine the best approach and landing technique to use. This may be an obstacle clearance procedure, a short field procedure, the procedure for a soft or rough field or some combination of these.

Don't let the pressure of the situation or preoccupation with the approach make you forget the pre-landing cockpit check.

Some Examples

The following are examples of how these procedures can be adapted to various situations. Notice that each features both an inspection of the proposed landing path as well as a normal circuit to prepare for a landing. Compare and contrast the examples, to understand how and why the procedures suggested differ. Try to think of situations in which you would use similar procedures or modify them further. Discuss these ideas with your instructor to obtain further suggestions.

Landing at a Familiar Aerodrome

Let's assume you are approaching a familiar airport, either your home airport or one from which you fly frequently. Also, assume that there is reason to suspect that the runway condition may have deteriorated somewhat since you last landed there. Perhaps it has snowed, or there is concern about recent rain, or there may be uncertainty as to whether or not recent construction work has been completed. Whatever the reason, you are justified in obtaining more information before you attempt to land.

Use whatever radio advisory services you can to determine the runway condition. If no service is available or you are still in doubt fly a normal circuit on to the final leg. Rather than land on the runway, fly level beside the runway as low as is safe and necessary to inspect the surface. If still unsure of the surface condition, make as many more passes as are needed. If satisfied with what you see, do another normal circuit and complete whatever type of landing observations tell you is appropriate. If observations tell you that the surface is unsuitable, depart the circuit in the appropriate manner and proceed to an alternate landing site.

Landing at an Unfamiliar Aerodrome

Again, use whatever radio advisory services you can to determine the runway condition. If no service is available or you are still in doubt, fly overhead the airport at least 500 feet above circuit altitude to determine which runway to use, circuit direction, and what features other than the runway could affect selection of a landing path.

Once you have decided upon the runway and circuit direction, proceed to the inactive side of the runway, descend to circuit altitude, and join the circuit. Once in the circuit, complete as many inspection passes as needed to determine the suitability of the selected landing area and decide on the type of landing. If you like what you see, complete one more circuit and land. If you don't like what you see, depart the circuit and go on to an alternate destination.

Landing on an Unprepared Surface

Although the surface on which you propose to land is unprepared, using standard circuit procedures will help you to orient yourself. As at an unfamiliar aerodrome, if possible, a pass well above circuit altitude has considerable benefit. It provides the opportunity to check for wind, traffic and obstacles. It also allows you to observe the general layout of the proposed landing site and see possible approach and departure paths that may not be obvious from lower altitudes.

When satisfied that there is no reason to avoid the landing site, descend on the inactive side of the "runway" set up a normal circuit pattern, and complete as many inspection passes as required to determine the site's suitability.

As in the other situations, keep looking until sure the location is acceptable. If you like what you see, complete another circuit for landing. If you don't like what you see, depart and locate another landing site.

Landing with Minimum Time for the Inspection

A good rule is to use as much time as you require to determine the information needed to plan a landing. However, make the decision to land before such things as deteriorating weather, approaching darkness, or low fuel make an abbreviated inspection procedure necessary.

In some situations, the time required for detailed inspection may not be available. Perhaps engine problems or illness on board dictates an immediate landing. In these cases you must find ways to abbreviate the procedure.

Perhaps the inspection passes above circuit altitude can be eliminated. Perhaps you can inspect the site adequately while passing overhead the field on crosswind. While on the inactive side of the landing area, you may be able to fly parallel to the landing path and observe the surface from that side and also from the downwind leg. Perhaps base leg or some other portion of the circuit affords an adequate look at the landing surface.

There are many ways to shorten the procedure. However, take sufficient time to obtain the information necessary to ensure a safe landing.

Making the Right Decision

Build on your experience and on that of other pilots. Each precautionary landing is unique in many respects. Each situation presents an opportunity to combine your pilot knowledge and skill with the ability to analyse the situation and make safe decisions.

Forced Landing

Engine failures are remarkably rare but they do happen. This fact makes it extremely important that you become proficient in the execution of forced landing procedures to bring you to a safe landing.

During earlier training you learned about gliding for range and how to estimate the point of touchdown by referring to the terrain ahead of you. You also learned how to determine the distance you can glide allowing for the effect of the wind. During practice landings, you learned how to choose a point from which a successful power-off approach and landing could be made. These skills are developed further in forced landing training and practice. It is strictly a matter of practical and methodical application of what you already know.

Those who believe that a successful forced landing is difficult to achieve are reminded that every landing made by glider pilots must be a successful forced landing, and they develop this skill very rapidly as the average student pilot reaches licensing standard in less than 8 hours total flight time.

Initial Actions

Should your engine fail totally or partially, follow the procedures recommended in the Aircraft Flight Manual. In the absence of manufacturer's instructions the following steps should be taken immediately and in order:

1. Control the aircraft — establish a glide, place carburettor heat on, and trim.
2. Select a landing site.
3. Plan the approach.

Control the Aircraft — Establish a Glide, Place Carburettor Heat On, and Trim

Establish the aircraft in a glide attitude at the speed recommended in the Aircraft Flight Manual. Once you have established the glide and placed the carburettor heat on, trim the aeroplane. Should the aeroplane not be properly trimmed, variations in gliding attitude will reduce gliding distance, making it difficult to judge accurately the approach and the touchdown point.

A common cause of engine failure on carburettor equipped aircraft is carburettor icing. If the engine fails totally, or if power decreases to a low value, engine heat will be lost rapidly. Therefore, it is imperative that you apply carburettor heat simultaneously with establishment of the glide to make use of the remaining engine heat. On aircraft with fuel injected engines the alternate air source to the engine should be opened at this point. These procedures may remedy the icing problem, if that is the cause, and prevent a forced landing.

Select a Landing Site

Always be on the look-out for suitable landing fields. Before selecting the landing site, determine the distance the aircraft will glide. Then look ahead, to either side, and if time and altitude permit, bank the aeroplane and take a look below and behind you. This is to ensure that no potentially good landing site or aerodrome has been overlooked. Fields may be non-existent in mountainous or heavily forested areas. Roads and highways may be your best choice in some cases, but beware of unseen hazards such as power lines and signs. Traffic must also be considered. Naturally the best field for a forced landing is an established airfield or some hard-packed, long, smooth field with no high obstacles, at least at the approach end. Since no guarantee can be given as to the location of your forced landing, you must learn to select the best available field. Cultivated fields are good, provided you remember to land parallel to the furrows, but fields used for pasturing animals usually harbour boulders or tree stumps. Avoid fields with contour

127

plowing, deep ditches, or with any other features that reduce the suitability. Try to pick a good field near houses, or at least near a road. This is particularly important in the winter.

When choosing a field, you must take it's length into account. If a strong wind is blowing, the normal landing roll will be comparatively short, but if it is imperative that you land downwind, the landing roll will be extended. Similarly, the existence of a slope affects the length of the landing roll. Normally you should always attempt to land into wind. If this is impossible because of lack of altitude, or the absence of a suitable field, carry out a crosswind landing, or as a last resort, a downwind landing.

Determining the Wind Direction

Nature provides numerous methods of determining the direction of the wind. Smoke gives the best indication. If smoke rises and drifts off slowly, the wind is light; if it rises and abruptly breaks away, the wind is probably strong. Grass and grain fields ripple with the wind, and dust blows with the wind. If it is impossible to tell from which direction the wind is blowing, land in the direction of the wind at the time of take-off.

Low-Altitude Engine Failures

Engine Failure on the Runway

If partial or complete failure is encountered on take-off, while the aircraft is still on the runway, close the throttle and apply the brakes. If it is obvious that the aircraft cannot be stopped before it runs off the runway, select the battery master switch and the fuel valve to their "Off" positions. Try to avoid obstructions such as fences or ditches, which may badly damage the aircraft.

Engine Failure After Take-Off

If the engine fails immediately after take-off, you may only have time to close the throttle, attain a recommended gliding speed, pick a landing path, and concentrate on a good landing. Do not become so engrossed in doing checks that you jeopardize the chances of making a good approach and landing.

Engine Failure Below Circuit Altitude

Engine failure while climbing after take-off gives you

a little more time to assess the situation. You should be able to select a field straight ahead; then carry out the following steps:

1. Close the throttle.
2. Lower the nose to maintain the glide speed.
3. Land straight ahead, or alter course slightly to avoid obstacles.
4. If time permits, complete the "Cause Check." Call "Mayday." Advise your passengers.
5. Secure the engine.
6. Carry out a forced landing.

Numerous fatal accidents have resulted from attempting to turn back and land on the runway or aerodrome following an engine failure after take-off. As altitude is at a premium, the tendency is to try to hold the nose of the aircraft up during the turn without consideration for airspeed and load factor. These actions may induce an abrupt spin entry. Experience and careful consideration of the following factors are essential to making a safe decision to execute a return to the aerodrome:

1. Altitude.
2. The glide ratio of the aircraft.
3. The length of the runway.
4. Wind strength/ground speed.
5. Experience of the pilot.
6. Pilot currency on type.

Should you have only partial power, it may be possible to complete a circuit and execute an emergency landing.

Engine Failure in the Circuit

When flying a normal circuit, it is highly probable that a forced landing can be completed successfully on the runway in use, or on one of the other runways more suited to your position. As soon as the engine fails, carry out a normal forced landing.

Forced Landing From Altitude

Circuit Forced Landing Pattern

When an engine failure occurs, judge the glide to approach and land as you did when you were practising power-off approaches and landings at your home base. A key position may be defined as some physical features on the ground that are chosen to provide continuing orientation to the field selected for forced landing purposes and to establish a near to normal base flown. Plan your glide to the key position to correspond as closely as possible to the familiar circuit pattern. However, depending on altitude and position, it may be necessary to fly directly to the key position, or fly a direct or straight in approach.

360° Forced Landing Pattern

There are some situations in which a 360° forced landing pattern (Fig. 2-57B) can be useful, even essential. This forced landing pattern is a gentle 360° descending turn. The turn is started with the aircraft heading in the direction of the intended landing at an appropriate altitude over the desired touchdown point. This starting position is called the *high key*. The appropriate altitude for this high key depends on the glide performance of the aircraft. To calculate this altitude, take the amount of altitude your aircraft normally loses in two minutes of gliding descent, add 200 feet, and you have the height above ground that works well for high key. For example, if your glide rate is 600 feet per minute, the high key altitude should be 1,400 feet AGL (2 x 600 + 200).

The 360° forced landing pattern is actually quite easy to execute, but it does require that you estimate the elevation of the landing area and can manoeuvre the aircraft successfully to the high key position. Once at the high key position, you start a gentle turn to reach the *low key* position, which is half-way around the circle and abeam the touchdown point. The aircraft at low key will be about a one-minute glide lower than it was at the high key, the aircraft will be heading downwind, and you will have a good view of the landing area on your wing-tip. From there, the turn is continued to the *final key*, where the aircraft will be around 500 feet AGL in a tight base leg position.

Like any forced landing pattern, things may not always work out perfectly. If you find it impossible to arrive at high key at the recommended altitude, *do not panic!* Instead, try to reach low key at the correct altitude. If you are really high when you reach the high key position, you can do a 360° orbiting turn down to the correct altitude. If you are just a few hundred feet high, fly upwind until you lose half the excess altitude. The remainder will be lost on the downwind as you turn

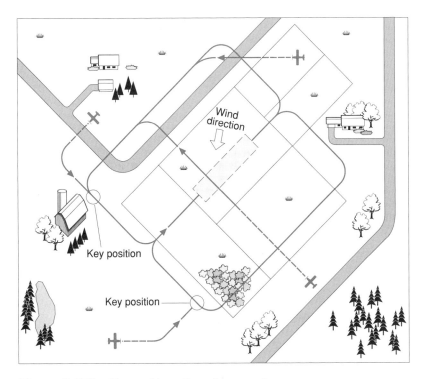

Figure 2-57A   Forced Landing Approach

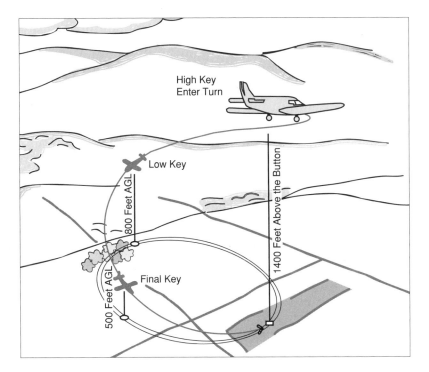

Figure 2-57B   360° Forced Landing Pattern

leg distance from that field. Distance in this case must be well within the into-wind gliding range to the field. In a strong wind, the base leg must be flown closer to the field than in a light wind. Key positions are shown in Fig. 2-57A. The flight paths shown are a few examples of how an approach to a forced landing might be the circle into a slight oval racetrack pattern. If you are slightly low when you reach high key, tighten the turn slightly, but not more than a medium turn. This will keep you closer to the landing area and allow you to get around to low and final key more quickly. If there is a strong wind, move the high key down the field to compensate for the way the wind will distort the pattern. If there is only a light wind, use slightly less bank during the first half of the turn and slightly more bank during the second half to maintain a good pattern. Exercise 20 explains drift control during ground reference manoeuvres.

One of the difficulties with the 360° forced landing pattern is most noticeable with low-wing aircraft. It can be difficult to know when you are directly over the touchdown point because the wings block the view below. For this reason, once you choose the landing area, look for roads or other landmarks that are abeam the high key position and can be used as references once you are overhead. Fortunately, the low key position is quite easy to identify with a low wing aircraft. As soon as you bank to start the turn, the low key position will be the point you see on the ground when you look down the lowered wing.

No matter what pattern is used, you should plan to arrive in the correct position for a normal final approach but slightly high. When a landing can be made into the first third of the selected site, use flaps or side-slip to lose excess altitude so that touchdown can be made at the most desirable point on the landing area. Remember that there are ways to lose altitude, but there is no way to correct an undershoot. *Never stretch a glide.* Practise forced landing often and vary the entry altitude, field, and geographical location. This will help you to develop the confidence, skill, and judgement necessary to carry out this procedure safely should a real forced landing become necessary.

Cause Check

Many actual forced landings need not have occurred, as the failure or near failure of the engine was caused by something within the pilot's ability to remedy. For this reason, forced landing training includes certain checks which should be carried out during the descent if time permits.

The actions that follow have no set time or place where they should be done. They are suitable for most light aircraft and may be used in the absence of manufacturer's data.

1. (a) Fuel on and amount. Check that the fuel selector is on the correct tank or has not been placed accidentally in the "off" position. If already selected to a tank indicating a good quantity of fuel, select another tank known to contain fuel. The gauge may be faulty or a line blocked.
(b) Fuel Pumps. Place the switch in the "on" position, as there may be a failure in the regular fuel feed system.
2. Primer Locked. Check that it is in and locked. A primer that has worked itself out through vibration can cause the engine to malfunction.
3. Mixture Rich. Check that the mixture control is in the "full rich" position; it could have been moved accidently into "idle cut-off" or left in "lean" during a descent.
4. All Switches On as Required. Check that the magneto switches are in the correct position. If the engine continues to run roughly, select left and right positions to determine if it will function smoothly on one magneto.

MAYDAY

Depending upon the geographical location, altitude, and situation, the cause check and mayday action may be transposed. As an example, it may be more important in remote areas to let someone know you are in trouble before you attempt to find a cause for the engine failure. If you wait too long to make an emergency transmission, you may have descended below the radio range of ground stations.

Communicate with someone. If already established on a specific frequency, inform that unit of your problem. If uncertain that you can make contact on any selected frequency due to the time and altitude remaining, switch immediately to the emergency frequency of 121.5 MHz and transmit to ALL STATIONS.

Remember that those receiving an emergency radio transmission will be anxious to help and will want to know as much about your situation as possible. You may have very little opportunity to spend time or effort answering their questions; therefore, your initial call should be organized and clearly spoken.

In the case of a forced landing, the word MAYDAY should be spoken three times. This is followed by your AIRCRAFT IDENTIFICATION three times and then the message. The message should briefly state your problem, geographic location, and intent. For example, engine failure over the town of "Sumwhere," landing in a field approximately two miles north of the town. Your aircraft identification should be repeated at the end. As well, if your aircraft is transponder equipped, ensure it is ON and select code 7700.

Passenger Safety

Advise passengers to remove glasses and all sharp objects and stow them securely with other loose objects.

In a two-place side-by-side aircraft, have the passenger move the seat as far back as it will go to assist in preventing head injuries. All seat-belts and shoulder harnesses should be fastened snugly. On final approach, provided the aeroplane is not affected aerodynamically, unlatch the doors to prevent them from jamming if the aeroplane fuselage suffers damage on landing.

Engine Shut-down

Shut down the engine to reduce the chance of fire. The fuel should be turned off, the mixture placed in idle cut-off, and the magnetos turned off. As well, the alternator or generator and master switches should be turned off, but the master switch should not go to off until the flaps are set for landing. It is important to perform these checks in the proper sequence. Therefore, you should refer to the appropriate section of the Aircraft Flight Manual for your aircraft.

Other Considerations

Close the throttle. An engine with a fuel flow malfunction may cease firing completely at a high throttle setting, but deliver enough power at a lower one to be of considerable value. Conversely, with the throttle still open, the engine may pick up because of a change in aircraft attitude just as you have made the landing site, only to fail again when you are carried beyond a safe landing. You might also try adjusting the mixture to see if a leaner setting will restore at least partial power.

Approach Height

The objective in making a forced landing is to have the aircraft touchdown at minimum safe speed at the most desirable point in the selected site. It is good practice to arrive on final approach with slightly more height than necessary so that you have more freedom of choice.

Another consideration is that in a real forced landing the gliding angle with the propeller windmilling or stopped may be steeper than during practice with the engine throttled back.

Simulated Engine Failure

When practising forced landings, the method used to simulate a power failure will be outlined by your flight instructor.

You must not allow the engine to get too cold or it may fail to respond properly when power is applied. Cruise power should periodically be applied for a few seconds during descent. This keeps the temperature normal and prevents the spark plugs from fouling due to prolonged idling of the engine. Many aircraft require application of carburettor heat before reducing power. Some aircraft can be configured with a specific power and flap setting combination that approximates a normal power-off glide; thereby, alleviating the need to apply power at intervals during the approach.

With the exception of those approaches made to an aerodrome, all simulated forced landings should be practised in the local practice area, and only to the minimum altitude specified by the *Canadian Aviation Regulations* or the training unit when they are more restrictive.

Pilot Navigation

One significant advantage that an aircraft has over most surface transportation is that it is capable of proceeding more or less directly to its destination at a constant and relatively high speed.

Pilot navigation is the most comprehensive of all the exercises you will learn, and some say the most satisfying. Piloting an aircraft cross-country requires all the knowledge and flying skills that you have acquired so far.

Planning and preparation is the key to a safe and effective cross-country flight. Effort expended before the flight can save a lot of effort and reduce stress when you are airborne. Once you take off you must be well organized and be able to perform in-flight calculations. You must be able to recognize features on the ground from the symbols on the map. You must be able to look out, listen out, carry out cockpit checks, assess your progress, revise estimates, fly accurately and, if necessary, divert to an alternate destination.

The exercise abounds in decision-making situations. You must be able to find out all you can about each situation, identify options for dealing with each, choose an option, act on your choice, and evaluate the results of your action. You should be aware of the factors that can lead to errors and apply the appropriate counter measures.

Initial Preparation

Preparation for a cross-country flight takes time, and the most important thing you can do is to make sure you allow adequate time. Doing some of the preparation before the day of the flight gives you an advantage, although this won't always be possible. Some of the things you can do beforehand include selecting your route, preparing the chart, checking aerodrome information and chart updating data in the *Canada Flight Supplement*, and entering the direction and distance of each leg of your route in the navigation log. You can also anticipate your needs on the day of the flight, make arrangements for a weather briefing, and advise passengers in advance to dress for the expected conditions. As well, give some thought to weight and balance at this stage. Make sure your passengers know how much baggage they can bring and if weight means you won't be able to carry full fuel then make arrangements for the correct amount of fuel to be in the tanks on the day of the flight.

Final preparation is done once you have received the necessary weather forecasts and reports on the day of the flight.

Selecting the Chart

For any cross-country flight you will need current maps in good condition. If your route takes you close to the edge of a map, you will want to have the adjoining map as well.

The standard chart used for pilot navigation in Canada is the 1:500,000 scale VFR Navigation Chart (VNC). The comparable chart used in the United States is the Sectional Aeronautical Chart. World Aeronautical Charts, with a 1:1,000,000 scale, can be useful for general planning of a long route, and some pilots carry them as another aid during flight, although they lack the detail needed for effective navigation at low and medium altitudes. To satisfy special operational requirements at certain high-density traffic airports having complex airspace structures, VFR Terminal Area (VTA) charts are available in a scale of 1:250,000. A list of airports for which these charts are published can be found in the MAP section of the *Aeronautical Information Publication* (A.I.P.) Canada.

Choosing the Route

Planning for a flight begins the moment you decide you want to fly somewhere. Deciding where you want to go and where you will start the trip means you are making important decisions about your route very early in the process. Exactly how you get from one end to the other requires that you make more decisions. A straight line from beginning to end is certainly the shortest, but compromises that result in a longer route are sometimes needed. Choose a route that offers both alternate airports and good landmarks. Note the elevation of terrain, giving particular attention to hills, peaks and other obstructions. A mountain range could block a direct path, or there might be a very large body of water you would be wise to go around. Airspace is another consideration. Restricted areas must be avoided and entry into these areas requires prior permission. There may be areas of busy traffic you would prefer to avoid, such as an advisory area or the airspace near a busy airport. If refuelling will be required, choose a route near airports that have the fuel you need at the time of day you will be there. It may also be wise to consider a more populated route, perhaps even a recommended VFR route such as you find in mountainous regions. All in all, there are many reasons, including weather, why your route may often involve several legs rather than one direct leg. Safety must always be a major consideration in the choice of a route.

Preparing the Chart

Before you draw lines on the chart, make sure you have the necessary tools at hand — pen, pencil, marker, protractor, dividers or chart rule, and a straight edge long enough for the sections of the route.

Track. The first line to be drawn on the chart is your intended track. The line should be dark and neat and easily distinguishable from other lines on the map. Some pilots like to emphasize their track lines using a highlighting marker, taking care not to obscure important detail.

Drift Lines. These lines are drawn at 10 degrees each side of the required track from both ends of each leg of the route. Extend the lines about two-thirds the length of the leg and make them distinct from the track line. One way to do this is to make them dashed lines.

Ten-Mile Marks. A small stroke across the track line at ten-mile increments can be very helpful in assessing your progress in flight.

Fractional Distance Marks. Dividing a leg into equal segments such as one-quarter, one-half, and three-quarters allows easy and rapid revision of ETA without using a flight computer.

Check Points. Although map reading proceeds continuously throughout a flight, many pilots like to choose distinct landmarks at ten to fifteen minute intervals along their track and identify them with arrows or circles. Take care when writing on the map to avoid obscuring detail. Although the markings suggested here are all useful, if they were all used on a very short leg the result would be confusing. In this situation, draw the track line and decide which markings would be most helpful on the leg.

Final Preparation

With the initial preparation done, the preparation on the day of the flight will be a lot easier.

NOTAM

Check NOTAM. This is where you will find out important information about aerodromes you will be using. NOTAM will tell you if certain facilities, runways, or airports will be closed. Look for notices that will affect your progress along the route. You might find that a navigation aid you were planning to use is shut down, or there might be radio frequency changes. You might see that a forest fire is raging at some point on the route and restrictions may apply to your direction of flight or altitude. There might be special areas designated for military flying or search and rescue operations might be in progress. NOTAM are important and they should be checked before you depart on a cross-country flight.

NOTAM are not provided for some aerodromes. If in doubt about field conditions, call ahead by telephone or radio to obtain current airport information. Telephone numbers are listed in the *Canada Flight Supplement*.

Weather

Adequate weather information is required for you to make meaningful decisions regarding the cross-country flight. Should the flight be delayed until the weather is more favourable? What altitude would be most suitable or efficient considering terrain height and obstructions? Is there a route around the weather? Many pilots set personal weather limits for themselves

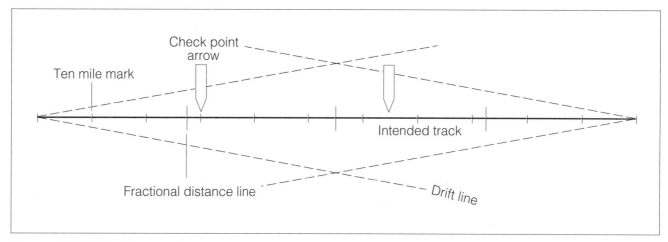

Figure 2-58   Chart Preparation

that are higher than the regulatory minima. If the forecast indicates the weather will be below these minima they don't go. In setting these limits, they take into account their experience, currency, familiarity with the route, terrain, alternate airports, and training.

The best way to obtain weather information is by a briefing from a specialist at a Flight Service Station or Atmospheric Environment Service weather office, either in person or by telephone.

You should brief the briefer. This will include identifying yourself as a pilot, giving your aircraft registration or pilot licence number, stating that your flight will be VFR, and giving the planned route, altitude, destination, departure time, and estimated time en route. The weather specialist will use this information to present an appropriate weather briefing for your planned flight.

When this information has been received and considered, the GO/NO GO decision will be yours to make. Don't forget to check the weather trend. This is done by comparing the current reports and two or three preceding weather reports with the forecast. Be alert for reports showing weather that is worse than what is forecast.

Choosing an Altitude

Care should be taken in choosing an altitude for your flight. The height of hills, peaks, and obstructions must be considered and your altitude should be appropriate for the direction of each leg of your flight. Weather must be considered, including ceilings, visibility, and upper winds. Distance should affect your choice of altitude. You won't want to climb to a high altitude if you are only going a short distance. Fuel economy, passenger comfort, and the effective range of any navigation aids that you might be using must be considered. If you are considering a very high alti-

tude, remember the requirements for oxygen and know the service ceiling of the aircraft. If you must cross a body of water, choose an altitude that will let you glide to the shoreline with some altitude to spare in the event of engine failure. Comply with any airspace restrictions that may limit your altitude. Think of your requirements for map reading. You can see more detail at lower altitudes, and you can see farther at high altitudes.

The Flight Planning Form

A flight planning form is an organized plan and log for the flight which minimizes the possibility of forgetting important data and having to do computations in flight (Fig. 2-59). There is no correct form, as different designs reflect different needs and preferences, but they all have to balance the need for complete information with the danger of clutter that can make the log difficult to use in flight. Many pilots prefer a log printed on 8 1/2" by 11" paper with one-half given to planning the trip and the other half to in-flight record keeping. Some training organizations like to provide space for weather, NOTAM, and weight and balance calculations. Others like to use separate sheets for this purpose.

The Flight Plan, Flight Itinerary

Telling someone where you are planning to go, and when and how you plan to get there is simply good sense. A *flight plan* filed with Air Traffic Services is the best way to do this. Pilots can also file a *flight itinerary*, which means giving the details of your flight to a responsible person who will notify appropriate authorities should you fail to arrive at your destination.

FLIGHT PLANNING FORM (✓)

From — To	Alt.	IAS	TAS	Track (T)	W/V	Hdg. (T)	Var.	Hdg. (M)	G/S	Dist.	Time	Fuel Req'd
							Distances/speeds in				kts. / mph	✓
MELFORT - CHECK PT #1	↗	90	–	135			17E	118		2	2	
CHECK PT #1 - HUMBOLDT	4500	115	120	204	290/20	214	17E	197	117	43	22	8
											24	8

Weather Forecast

YXE/YPA

SCT 050 OVC 070 6-8 SM -SHRA
OCNL GUSTS 30
3-280/15   10
6- 300/25   6

En Route Station Reports

En Route Radio Frequencies and Navigation Aids

YPA   VOR 113.0
YXE   VOR 116.2
      TWR 118.3

315' TOWER 1 MILE N OF HUMBOLDT

Destination Information

YXE SCT 040 OCNL 6 SM -SHRA

Forecast W/V 275/20G25

Runway   09-27 1900' TURF
          1865'ASL

Cross-wind component   10° 6KT.
HUMBOLDT- 072°RADIAL YXE VOR
PHONE YXE FOR WX 306-665-4265
FLIGHT PLAN YXE 306-242-8227

Figure 2-59   Sample Flight Planning Form

Whether you use a flight plan or a flight itinerary, the aim is to have someone find you as quickly as possible if something goes wrong, and you are forced down.

Departure

There are three basic departure procedures you can use when starting a cross-country flight. One is the *overhead departure* where you climb in the vicinity of the airport to reach cruising altitude and then set heading overhead the airport. This takes time and fuel and can conflict with other traffic, but it does give you a known fix for starting. It is often used when the airport is surrounded by miles of relatively featureless terrain or the visibility is limited. Another departure is the *geographic point procedure* in which you choose to set heading at a distinct landmark a short distance away. This method allows you to concentrate on departure procedures before you become concerned about setting heading. It is often used when departing from busy airports or those where airspace restrictions apply. Be careful to choose a geographic point that is distinct and easy to find. The last type of departure is the *en route climb*, in which you take off and at a safe altitude turn to intercept track and climb en route. This is quick and direct, but you can become very busy and it is sometimes difficult to determine when you are established on track.

Level-off Procedure

Level the aircraft at the altitude you want and at the speed you planned and trim carefully. Check the outside air temperature to confirm that it is what you expected. If it isn't, and you planned to fly a particular indicated airspeed, your planned true airspeed won't be correct.

Lean the mixture if required and set the heading indicator. Scan your aircraft engine and systems instruments and check your fuel selector and quantity.

Setting Heading

It is useful to have a routine or procedure you can use at the beginning of any leg of a cross-country flight. When established over a set-heading point on the desired compass heading, record the time as this is the basis for all your estimates and ETAS. Maintain your heading and check the *visual angle of departure*, compare the heading indicator and compass and check major features on the map to confirm that the aircraft is in fact heading in the correct direction. Finally, calculate an estimate for the first check-point and an ETA for the end of the leg.

En Route

Map Reading

Arrange charts so that both the charts and the controls of the aircraft may be easily managed at the same time. A chart should be folded so that the section being used is readily available, with a minimum of refolding or handling in the air. If more than one chart is to be used, they should be pre-arranged in the order in which they will be required.

Effective map reading can be broken down into four steps: orientation, anticipation, confirmation, and pin-pointing.

Orientation of the map means holding it so that your map track parallels your ground track. This makes features on the map easier to compare with features on the ground, which is far more important than holding the map so you can read place names.

Anticipation means using the one instrument that equals the compass in importance in navigation — the watch. The expression "watch to map to ground" means to first look at your watch to anticipate features that should come into view in the next few minutes. By anticipating where you are going to be you can study the map to familiarize yourself with the features that will enable you to positively identify a place on the ground. This technique can change to "watch to ground to map" if you are uncertain of your position.

Confirmation means taking care to positively identify a landmark. For example, if it is a town, make sure all the features around it — roads, railroads, streams, lakes, power lines, and other details — relate correctly to confirm your identification.

Pin-pointing means identifying your position relative to a time and place and noting this on your map. Many pilots like to mark a dot at their precise position, circle the dot and write the time beside it.

Ground Speed Check

For a ground speed check to be accurate, you should be established at your cruise altitude, heading and airspeed for the entire distance of the check. A ground speed check should be done at the first positive pin-point, ideally between 10 and 25 miles from the departure point. Use the flight computer unless the values permit easy mental calculation. Further ground-speed checks and ETA revisions should be made at any time when conditions such as track, wind, and airspeed change.

Cockpit Checks

Cockpit checks should be carried out at regular intervals. Include such items as setting the heading indicator, checking aircraft engine and systems instruments, and confirming that fuel is being consumed at the rate you expected.

Position Reports

VFR position reports are easy to give and should be a routine part of every cross-country flight. Updating your position lets others know where you are at a given time, and should your aircraft become overdue, it can significantly reduce the time it takes Search and Rescue to locate you.

Pilot Weather Report (PIREP)

Reporting weather conditions you encounter en route can be very helpful to other pilots. Any conditions differing substantially from those indicated in forecasts or reports will be important. Pilots will be particularly interested to hear about any turbulence, icing, thunderstorms, strong winds, heavy precipitation, or reduced ceilings and visibility. PIREP can be passed to any Air Traffic Services facility, such as a Flight Service Station.

Track Errors and Corrections

Owing to inaccuracies in the forecast wind, in navigational technique, and in flying the aircraft, errors often occur that require alterations of heading to bring the aircraft back on track. Before corrections are discussed, the following terms must be defined:

Required Track. The proposed path of the aircraft over the ground.

Track Made Good. The actual path of the aircraft over the ground.

Track Error. The angle between the required track and the track made good, measured in degrees, and always expressed as being left or right of the required track.

Opening Angle. The angle between the required track and the track made good.

Closing Angle. The angle between the old required track and the new required track to arrive at the destination.

Ten Degree Drift Lines

In Fig. 2-60, 10 degree drift lines are shown opening from the set-heading point and closing to the destination. These lines enable the pilot to estimate track errors and required heading changes with reasonable accuracy.

Point A in Fig. 2-60 indicates a point on the track made good that is 7 degrees left of the required track, indicating an opening angle of 7 degrees. The angle can be estimated more accurately by physically making a mark at mid-angle (line B) and establishing the pin-point in relation to that mark. Closing angles may be determined in the same manner, using the 10 degree drift lines that converge on the destination. The angle between the required track and a line joining

point C to destination indicates a closing angle of 4 degrees.

Heading Corrections

Once you have established the position of the aircraft accurately, and provided it is not on the required track, decide upon the best course of action. Normally 10 to 25 miles should be flown before attempting to estimate any track error, because errors over a short distance are magnified considerably. The heading of the aircraft can be changed to either return to the required track or fly directly to the destination. Your choice will depend on the position of the aircraft. It is usually more desirable to return to the required track, as that is the track you have studied and, in addition, a line already drawn on a chart is much easier to follow.

Provided the aircraft has not passed the half-way point, the double the track error method or the visual alteration method can be used. If the aircraft is beyond the half-way point, you can still use the visual alteration method, or alternatively, the sum of the opening and closing angles method. Each method will be discussed in detail later in this text but first some basic geometry may help.

The key to understanding how these various methods work lies in the fact that if the aircraft heading is altered in the direction of the required track by a number of degrees equal to the track error or opening angle, the resulting heading will produce a track parallel to the required track. Using the 10 degree drift lines in Fig. 2-61, it can be seen that the track made good is widening from the required track by an angle of 7 degrees. By altering the heading 7 degrees to the right, you now have the heading that produces a track parallel to the required track and is approximately the heading to steer when the required track is regained. What you now want is to regain the required track

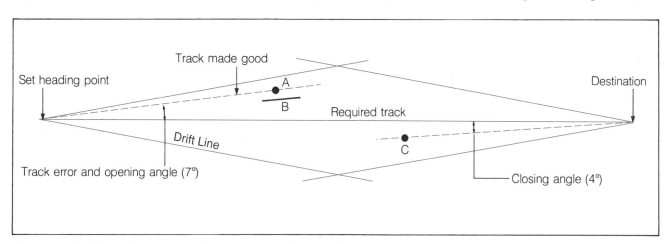

Figure 2-60   Track Errors and Opening and Closing Angles

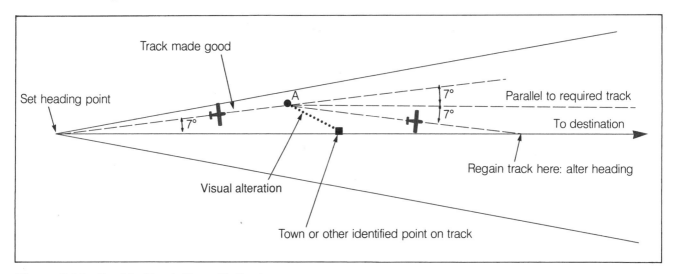

Figure 2-61   Double Track Error Method

using an orderly method. Once this is accomplished you can fly a corrected heading to maintain track. Alternatively, you have the choice of flying direct from the position off track to the destination.

As mentioned above, there are several methods used by pilots to correct track errors. When choosing one of the following, consideration should be given to visibility, availability of good landmarks, distance to destination, and pilot experience.

The Visual Alteration Method

From time to time, a pin-point will show you to be left or right of your required track, and one way to regain track is by *visual alteration*. Using Fig. 2-62 as an example, you establish your position at point A and note that the track error is 6 degrees right of your required track. Fly visually from point A to the positively identified pin-point on your required track. Upon reaching that point, a compass heading of 084

degrees (090 degrees −6 degrees) should be flown to remain on the required track. This method works best if there are good landmarks, and you are very sure that the chosen point is, in fact, on your required track.

The Double the Track Error Method

Should lack of landmarks or some other reason preclude use of the visual alteration method, you should use the double the track error method. Simply double the amount of track error or opening angle, and apply this to the original heading in the direction of the required track: the aircraft will regain track in approximately the same period of time as it took to drift off track. The required track will be intercepted again at a distance along the track equal to twice the distance from the set-heading point to the point where the heading change is made (point D in Fig. 2-63). On regaining track it is necessary to subtract half the

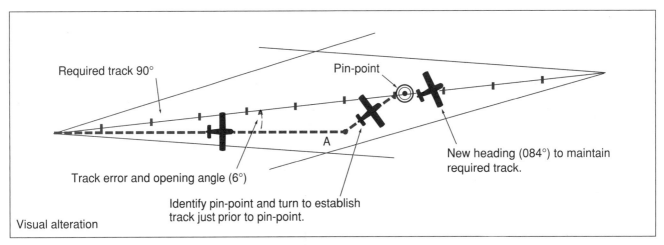

Figure 2-62   Visual Alterations

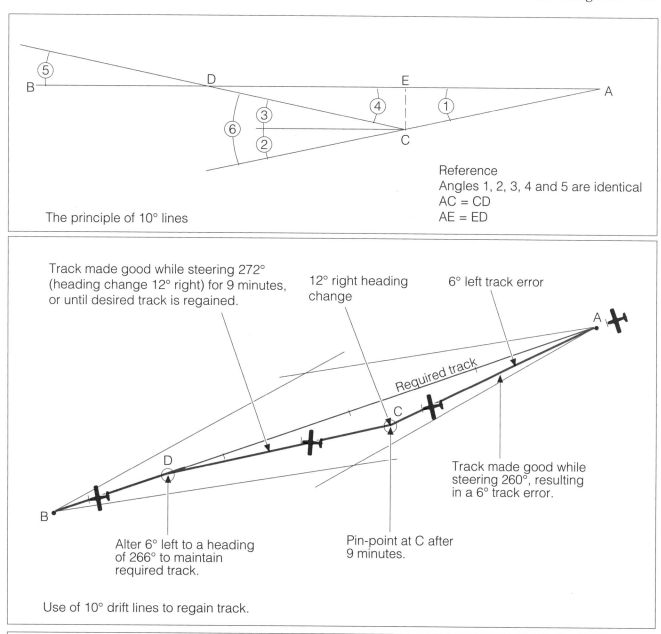

The principle of 10° lines

Reference
Angles 1, 2, 3, 4 and 5 are identical
AC = CD
AE = ED

Track made good while steering 272°
(heading change 12° right) for 9 minutes,
or until desired track is regained.

12° right heading
change

6° left track error

Required track

Track made good while
steering 260°, resulting
in a 6° track error.

Alter 6° left to a heading
of 266° to maintain
required track.

Pin-point at C after
9 minutes.

Use of 10° drift lines to regain track.

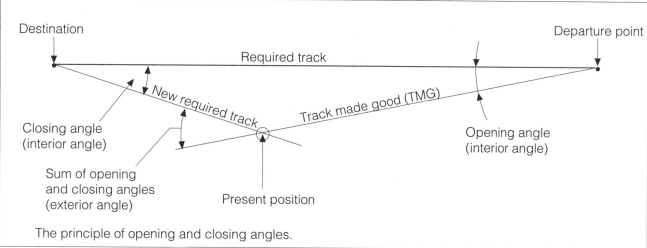

Destination

Departure point

Required track

New required track

Track made good (TMG)

Closing angle
(interior angle)

Opening angle
(interior angle)

Sum of opening
and closing angles
(exterior angle)

Present position

The principle of opening and closing angles.

Figure 2-63   Method of Making Track Corrections

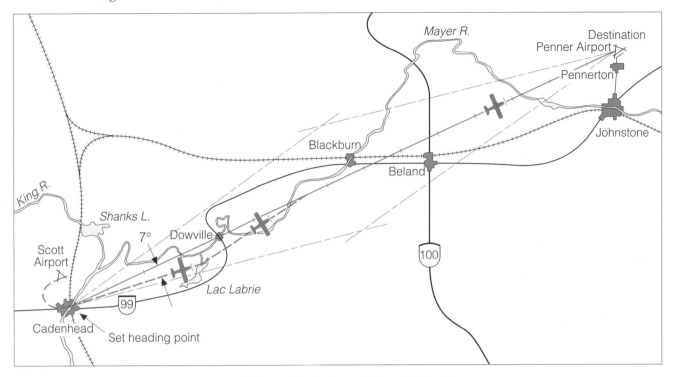

Figure 2-64   Tracking by Double Track Error Method and Visual Alterations

correction applied to the original heading to obtain the heading to keep the aircraft over the required track.

Now let's look at a practical example using Fig. 2-64. Take-off is from Scott airport and you set heading at 1000 hours over the point where the railway line crosses the King River, at the town of Cadenhead. Calculations indicate that you must steer a compass heading 065 degrees on the trip to Penner Airport, 148 miles distant. After 14 minutes (1014 hours) you pinpoint yourself at the north end of Lac Labrie, and using the 10 degree drift lines determine that you are 7 degrees right of track. A heading change of 7 degrees (058 degrees) left would result in a heading whereby you would parallel the required track. But as you wish to regain track using the double the track error method, you alter heading to 051 degrees (065 degrees − 14 degrees = 051 degrees). This new heading is held for an additional 14 minutes and you regain track just south of the bend in the Mayer River at 1028. At this time you alter heading 7 degrees right and steer 058 degrees. Any physical features near or along the track line that will confirm that the track has been regained will be helpful, but if none are available, the heading should be altered at the calculated time. Revised ETAS can be made while flying toward the required track.

The Opening and Closing Angle Method

Doubling the track error doesn't work once you are beyond the half-way point of a leg. Because it takes just as far to regain track as it did to get off in the first place, you would only regain track somewhere past your destination or turning point. The *visual alteration* or *opening and closing angle* methods will work in this case. As we have already discussed the visual alteration method, let's have a look at the principle behind the opening and closing angle method. Go back a few pages and study the definitions of opening angle and closing angle. By going back to basic geometry again and consulting Fig. 2-65, it can be seen that by altering the heading by an amount equal to the *sum* of the opening and closing angles, we should track to destination.

In Fig. 2-66 the line A-B passes through point C, where the heading change is made, and is parallel to the required track. At point C, which is 4 degrees left of track, a 4 degree right heading change would parallel the track. Angle X equals angle Y (the closing angle) and using the 10 degree drift lines we can determine that angle Y equals 7 degrees. Therefore, at point C, an 11 degree right heading change (4 degrees + 7 degrees = 11 degrees) should permit the aircraft to track directly to destination.

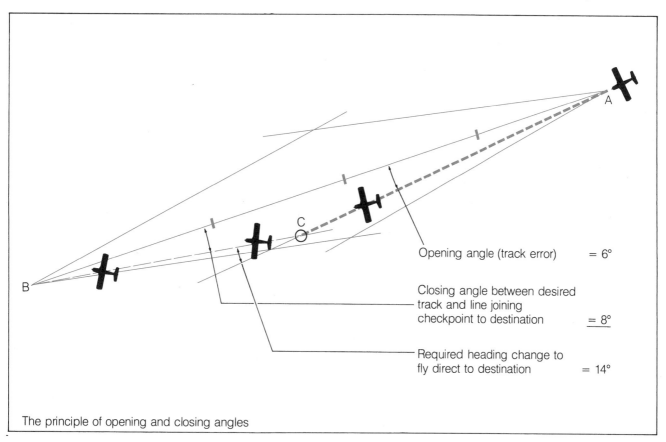

The principle of opening and closing angles

Figure 2-65 The Principle of Opening and Closing Angles

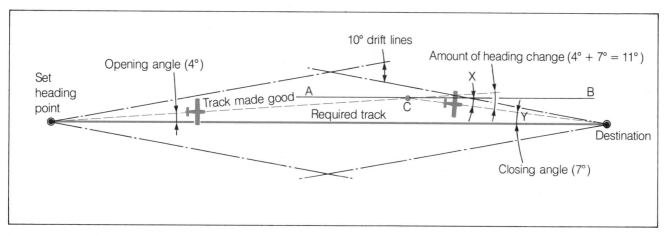

Figure 2-66 Tracking

Now let's look at a practical application. In Fig. 2-67 the calculated heading to fly from Speer Airport to Hemming Airport was found to be 105 degrees compass. Set-heading time was 0900. No reliable pinpoint was available until 35 minutes later, close to where the north-south railway crosses Highway 10, 82 miles along the track. Using the 10 degree drift lines, we find the opening angle to be 4 degrees and the closing angle to be 6 degrees. Therefore, to fly directly to Hemming Airport, a 10 degree right heading change (4 degrees + 6 degrees = 10 degrees) to a heading of 115 degrees (105 + 10 degrees) is required. As the ground speed is found to be 141 KT, and there are 49 miles to go, the revised ETA would be 0956. It should be understood at this time that the opening and closing method can be employed at any

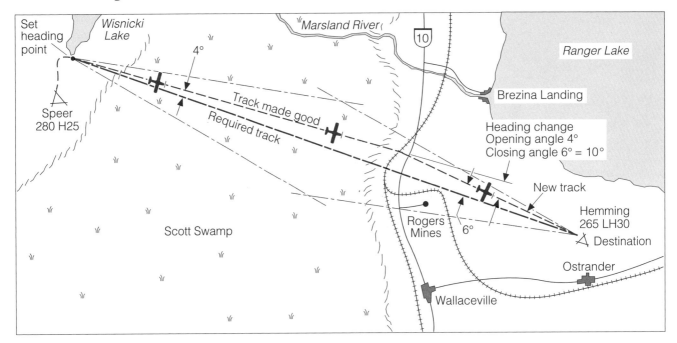

Figure 2-67 Tracking: A Practical Application

distance along the required track and is not limited to use after passing the half-way point. Alternatively, a visual correction could have been made to the point where the original required track meets the curve in the railway line at Rogers Mines, where a heading of 109 degrees (105 degrees + 4 degrees opening angle) taken up at the time would have kept the aircraft on the track to Hemming Airport.

The Drift Compensation Method

Experienced pilots often establish heading by constantly compensating for drift as the flight progresses. This is called the *drift compensation method*. When the heading to maintain a desired track is established by this method, drift angle can be easily computed if required. Visibility must be such that enough check-points are visible to maintain an accurate track.

Select at least two prominent check-points a sufficient distance apart (5 to 10 miles or more, depending on the terrain), on the desired track ahead of the aircraft. Maintain a heading that keeps the nearer check-point aligned with the farther one. When the aircraft and the check-points remain on the same line, the heading indicated on the compass will be the heading that, provided there is no wind change, should keep the aircraft on the desired track for the remainder of the flight. To continually compensate for wind effect, select another check-point on track in the distance before the nearer check-point is reached; then repeat the alignment and drift compensation procedure. Cal-

culate ground speed by time and distance between check-points to keep ETAs accurate.

Log Keeping

The pilot navigator has little time for log keeping in the air, but the following are some of the items that should be recorded:

1. Time off and flight plan opening time.
2. Set heading time.
3. Each compass heading (use 3 digits — e.g. 037).
4. The time setting each new compass heading.
5. ETA at planned turning points and destination.
6. Time over or by check-points and turning points.
7. New ground speeds and revised ETAs.

Other Considerations

Avoiding A Collision

The following are some of the simple strategies you can use to minimize the risk of collision with another aircraft. Check your physical, mental, and emotional condition and be on guard for absent-mindedness and distraction — two enemies of concentration during flight. Study your route carefully, looking particularly for areas of greater air traffic activity. Be particularly careful when you cross airways or enter areas where traffic is concentrated, such as advisory areas, and radio navigation aid sites. Keep the windows clean and clear of obstructions. Maintain proper separation from cloud. Follow the correct procedures for departing and joining the circuit. Minimize the time spent with your head in the cockpit by being well organized prior to the flight.

All aircraft have blind spots. A high wing aircraft has a critical one in the direction of the turn as soon as bank is applied. A low wing obscures the area beneath the aircraft. The most dangerous combination of events that could result in a collision arises when a low wing aircraft descends from the rear toward a high wing aircraft, or a high wing aircraft climbs from the rear toward a low wing aircraft. Therefore, in prolonged climbs and descents you should make small turns to ensure the area above and below is clear of aircraft.

Use your equipment. Aircraft exterior lights can help avoid collisions. Anti-collision, navigation, landing, and high intensity strobe lights will make your aircraft more visible by day and much more so by night. Transponders allow controllers to identify your aircraft's position in relation to other traffic.

Be sure to monitor appropriate radio frequencies when you are en route. This will help keep you informed about other traffic and weather. The *Canada Flight Supplement* specifies the frequency to use depending on the airspace in which you are flying. If you have two radios, consider listening out on the emergency frequency 121.5 as well.

Listen carefully to all radio transmissions. Monitor transmissions from pilots reporting their position to the tower, Flight Service Stations, and on published traffic frequencies. It will help you form a mental picture of traffic in the vicinity. Call at appropriate points to report your position, altitude, and intentions. As well, keep a good look-out as there may be aircraft nearby without radios or operating on a different frequency.

You must be able to scan a large area of sky in an organized way without missing small objects. Even the most perfect eye will not detect other aircraft unless the pilot is alert, expectant, and knows how to scan the sky. Identifying a target requires picking it up, locking on, and recognition. Moving targets are easier to see than stationary targets, and there may be other clues such as reflection and flashing lights. Unfortunately, aircraft on a collision course appear to remain stationary on the windshield until the last few seconds and so are more difficult to detect. Once you do sight the other traffic, don't forget the rest of the sky. If the traffic seems to be moving on the windshield, you're probably not on a collision course but continue to monitor its position. If the aircraft does not appear to have motion, then you should watch it carefully and be prepared to get out of the way.

Scanning is made up of two parts: the slow movement of the eyes across the field allows peripheral vision to detect a target or movement; then central vision is used for identification. There are different ways to scan but a common method is sector scanning. For example, the pilot divides the sky into approximately 30 degree segments and, starting in the middle of the windshield, scans slowly across each segment, pausing in the centre of each. The scan is carried over the two segments on the left and then brought back sharply to begin scanning the right-hand segments. Remember, the scan must not only be across a full field of vision (approximately 150 degrees) but must also cover an area 10 degrees above and below the horizontal plane. This will cover the areas most likely to contain significant traffic.

The most important tool for avoiding collision is your scan. Watch where you are going, and watch for other traffic. Use your scan constantly.

A good scanning technique takes time and effort to develop and must become a habit in all your flying. Keep in mind that it is an activity shared with other piloting tasks, and it can be degraded by fatigue, boredom, illness, anxiety, or preoccupation.

Two light aircraft approaching head on, with an airspeed of 120 KT each, have a closing speed of 240 KT. If they become visible to each other at the 3 mile range, they will have only 45 seconds to determine and execute the appropriate avoiding action. With a visibility of only 1 mile, the period of time in which decision and action must be taken is only 15 seconds. These times are based on a pilot-in-command actually sighting another aircraft at the exact point of visibility; in normal practice another aircraft may not be sighted until well into visibility range, which further reduces the time available for whatever action may be necessary.

Advice of Conflicting Traffic

In controlled airspace where air traffic is monitored by radar, Air Traffic Control (ATC) may advise when aircraft appear to be getting too close to each other by using the clock position system outlined in the *Aeronautical Information Publication*. When advised of other traffic, acknowledge and advise the controller whether or not it has been sighted. ATC may continue with advice until the traffic is sighted or a risk of collision no longer exists.

Arrival

Preparation for arrival at an airport should begin in the planning stages of the flight. Review the information given in the *Canada Flight Supplement* to familiarize yourself with the airport elevation, circuit procedures, frequencies, runway layout, and anything else you might need for a safe arrival.

Anticipate the destination information you will need as the arrival phase of your flight can be very busy, and it happens at the end of the flight when you might be tired. Details such as weather, altimeter setting, surface wind, runway in use, and other traffic will be important. If there is a control tower or a Flight Service Station then a radio call will be required. If ATIS is available, use it. Listening to other aircraft can tell you a lot, so tune in the destination frequency early and pay attention. Anticipate the type of circuit joining you will do and don't hesitate to position the aircraft directly overhead or ask for help if you are unfamiliar with the local landmarks.

Be sure to plan your descent. For example, if you have to lose 6,000 feet and you want to descend at 500 feet per minute for passenger comfort this will take you 12 minutes. If you descend at 120 KT (2 miles a minute), the descent will have to be started 24 NM early.

Be sure to file an arrival report, or if you have filed a flight itinerary, notify the chosen responsible person that you have arrived safely. You should also sign the arrival and departure register at uncontrolled aerodromes.

Diverting to an Alternate Destination

Knowing how to divert to an alternate destination is an essential skill. There are many reasons why a diversion may be necessary, for example, deteriorating weather, a sick passenger, or problems with the aircraft. Sometimes, of course, you may just decide you want to go elsewhere.

Making the decision to divert is often the most difficult step. You started with a particular destination in mind, and it may be difficult to accept the fact that you must now go elsewhere. It is a decision some pilots make too late.

Once you do make the decision, planning the flight to the alternate destination while continuing to fly the aircraft is a challenge that calls on skills you have already acquired. It may help to slow down while doing your in-flight planning, but keep in mind safe minimum operating speeds for your aircraft. Don't get so absorbed in planning that you forget your first task — fly the aircraft!

Choose a suitable destination within a suitable distance over suitable terrain. If there is a line feature such as a road, powerline, railroad, river or shoreline that leads to the new destination then use it. This method is the simplest and recommended if it is available. If there is no convenient line feature then draw a free-hand track line. If the diversion is particularly long, break it into shorter, more manageable legs.

Estimate Track

Estimating the magnetic direction of the track line is the next step, and there are various ways to do this without a protractor. You can compare your track line to other lines on the map, such as airways or track lines that you have used before, or you can hold a pencil over your line and move the pencil at the same angle to a VOR compass rose. These will give magnetic track. True direction can be obtained by estimating the angle at which your track line intersects lines of latitude and longitude. You must then apply variation to obtain magnetic track.

Reciprocal Track

One very common type of diversion is to return to your departure point. Estimating the heading that will maintain this reciprocal track is not difficult. First work out the reciprocal of your outbound track by adding or subtracting 180 degrees. If this is difficult, then either add 200 and subtract 20 or subtract 200 and add 20. Drift will be the same amount as you encountered outbound, but it will act in the opposite direction. Ten degrees left drift outbound will mean ten degrees right drift on the return flight.

Estimate Distance

There are a number of ways to estimate distance. The lines of latitude shown on your map are a convenient reference, since one minute of latitude equals one NM. If you prepared your original track line with 10 mile marks, you will also have these as a reference. As well, the compass rose circles around VOR stations are all approximately 30 NM in diameter on Canadian 1:500,000 aeronautical charts. Some pilots use the thumb and index finger as "dividers" and measure this span against a known scale. Others use the span of their knuckles and some use a pencil marked in 10 mile increments. It's surprising how accurate you can be using these methods.

Estimate Ground Speed

To estimate ground speed, you must have some idea of the wind velocity. If you received a weather briefing before your flight then you should have this information. Otherwise, you will have to estimate the wind by observing drift and ground speed. If there is no wind then ground speed and TAS may be assumed to be the same. If there is a direct tail wind or a direct head wind then the wind speed can be applied directly to the TAS, either adding or subtracting the amount. For example, if you fly at 100 KT TAS into a 20 KT head wind then your ground speed will be 80 KT. If the wind is coming from your wing tip position, assume no appreciable effect on ground speed. If it is coming from around 45 degrees to your direction of flight, assume a head or tail wind component of approximately two-thirds the wind speed.

Estimate Time

Ground speed is used to calculate time en route to the destination and any intermediate points you may find useful. This time estimate must be done mentally and is less difficult when put in the simplest possible terms. For example, a ground speed of 60 KT means time equals distance (one mile per minute) — 30 NM flown at 60 KT will take 30 minutes. If you fly at 120 KT (two miles per minute) time will be one-half the distance — 15 minutes in the above example. At 90 KT (1.5 miles per minute) then time will be two-thirds the distance or 20 minutes. You'll likely be close to one of these speeds and you can add or subtract a few minutes as required.

Another way to calculate a time estimate involves even less mental arithmetic. Divide the leg into equal parts and the time to fly the first segment is multiplied by the number of remaining segments. For example, if you divide the leg into four parts and fly the first part in six minutes, the remaining three parts of the leg will take 6 × 3 = 18 minutes.

Estimate Heading

Many pilots prefer to assume that track and heading are equal and upon setting heading they use the drift compensation method to establish a drift correction. This is quick, easy, and reliable. Alternately, if you know the wind is strong and anticipate a certain amount of drift, you can estimate and apply a drift correction at the outset and make adjustments if necessary.

Set Heading

Start from a known fix and note the time you set heading. Turn to an accurate compass heading to maintain track, reset the heading indicator, check your departure angle, and work out an estimated time of arrival for the end of the leg.

Advise Others

Once you have made the decision to go to an alternate destination and worked out the details, be sure to let others know. If you were on a flight plan, Air Traffic Control will need to know. Otherwise, you may find a search and rescue effort mounted when you are safely on the ground at some place other than where you said you would be. Worse, should a precautionary or forced landing become necessary, search efforts will be concentrated along your flight planned route, not your alternate route.

Low Level Navigation

If a diversion is necessary because of deteriorating weather, you may have to navigate at lower altitudes than normal. The greatest difference in navigating at cross-country altitudes lower than normal is the restricted field of view at low level, especially when flying over rough or hilly terrain. This, combined with the greater attention that must be given to handling the aircraft, reduces the time you have to positively identify your landmarks and check-points as they come up. As it is difficult to continually compare your check-points with the chart to assist identification, you must pick unique, easily recognizable features. Line features such as railway lines or roads that cross the track may be used as check-points, or followed if they parallel your track and lead to a turning point or to your destination.

Some landmarks that are easy to see at higher altitudes can be very difficult to recognize from low altitudes. For example, a lake might be hidden by a low hill. However, provided the visibility is not too limited, some features such as a radio tower might be easier to see against the horizon.

When flying over water, take special care to remain within gliding distance of land. It may also be a good idea to follow shorelines in winter, to reduce the risk of losing visual orientation because of white-out conditions (see Chapter 6, Weather Considerations).

Be alert and keep your head out of the cockpit as much as possible. Be on guard for rising terrain. Pay close attention to contour lines, spot heights, and obstacles along your track. Log keeping should be kept to essential items. When making log entries, check outside frequently if altitude is critical, to avoid flying into the ground. If you become uncertain of your position, climb as high as possible to give yourself an extended field of view and try to identify a landmark.

Despite your best efforts you can find yourself in the middle of a diversion that isn't going well. Be prepared to re-evaluate your situation and consider another course of action. If, for example, the weather is getting much worse, it may be necessary to abandon all thought of continuing a diversion and carry out a precautionary landing.

When approaching your destination, take care not to interfere with other traffic. If your destination is within a Control Zone, remember the requirements and procedures to enter a zone. If it is an uncontrolled airport, use the correct circuit joining procedure.

If You Are Uncertain of Your Position

There are times when you will be uncertain of your position. It can be unsettling to look around and not recognize any landmarks, but it certainly is no cause for alarm. It is a situation every pilot encounters, and such moments require calm reasoning and a recognized procedure. Taking slow, deep breaths may help you to think clearly.

Hold a steady heading and check the heading indicator against the compass. An error here can put you off track very quickly. Check your navigation log for possible errors such as drift correction or variation applied the wrong way. Make sure you are not using true heading or magnetic heading instead of compass heading. Check for a possible wind shift. Your log may show left drift when you can plainly see the aircraft is drifting right. If you do find an error, you can estimate where the error would likely have taken you and take steps to identify your position. If you are still close to your last known position, consider returning to this point. To do this you want to have very good landmarks behind you. If, after checking your heading and reviewing your navigation you still can't determine your position, try drawing a circle of uncertainty.

Circle of Uncertainty

When you are uncertain of your position, the normal map reading technique of "watch to map to ground"

changes to "watch to ground to map" which means you look for something recognizable on the ground and try to find it on the map. A circle of uncertainty will help narrow your search. The centre of the circle will be on your intended track line (unless you know your track made good) with the radius of the circle being 10 percent of the estimated distance flown since the last confirmed position. For example, if you have flown 10 minutes at 120 KT since your last known position, you will have covered 20 NM (2 miles a minute). Draw a circle 20 miles along the track from the last known position and make the radius 2 NM. Look in this area for landmarks. Remember, positive identification of a distinctive landmark is the only way to get back on track.

Manage Your Resources

If a circle of uncertainty doesn't fix your position, work your way step-by-step through a process designed to resolve the uncertainty. Do not exhaust your fuel in aimless wandering from one heading to another trying to pick up a landmark, and don't be afraid to admit to yourself and to others that you are lost.

Fly toward a major feature, such as a coastline, a railroad, or a highway. As you proceed toward it remember that once you get there you will have to decide which way to turn.

Climb if possible. This will help you to see farther and it will also increase your radio range. If radio contact is established, transmit your general position, amount of fuel remaining, request whatever assistance you need, and indicate the action you propose to take. Radar assistance or DF steers are available from ATC in many areas, and a heading to fly will be offered. If you have ADF or VOR and know how to use them, get a line of position or a position fix.

In an extreme emergency you should broadcast a MAYDAY message on 121.5 MHz. and listen out on the same frequency for instructions. If you have no idea of the direction to fly, set up a triangular pattern at endurance power settings, at the highest practical altitude, to alert the radar network (see *Aeronautical Information Publication* A.I.P. Canada, RAC). As well, if your aircraft is transponder equipped, ensure it is on and select code 7700.

Weather is often a factor in situations where pilots become lost, and flying aimlessly about in poor weather is a recipe for disaster. If you can't resolve the uncertainty, you must consider the possibility of a precautionary landing. You certainly don't want to fly until you run out of fuel or risk flying into an obstacle.

Instrument Flying

Instrument flight training is a part of private and commercial pilot training in Canada. An increasing number of general aviation aircraft have a full panel of flight instruments. More and more licensed pilots are choosing to broaden their competence by learning to control an aeroplane by reference to flight instruments alone.

While flying with reference to instruments, the control inputs required to produce a given movement are the same as those used in visual flight. The aircraft responds to the controls exactly as it has always done. There is a need to relax and apply control pressures smoothly, making small corrections and waiting for the results. The need for power and attitude changes must be anticipated to arrive at desired airspeeds and altitudes while accelerating, decelerating, climbing, descending, levelling off, and turning.

When you started flight training, you became familiar with the flight instrument indications associated with various attitudes and movements and saw how these indications compared to outside visual references. You learned to recognize instrument indications that related to the various attitudes and power settings used to control the aircraft. As well, you became aware of the fact that changes to attitude and power resulted in predictable changes to the instrument indications.

Of all the senses, vision is the one we rely upon most. However, if our normal visual flight references are taken away, we suddenly become prone to believing other senses that can cause confusion. Not being able to see the aircraft's position in relation to the ground may cause you to lose track of direction and attitude. This can lead to one of a number of illusions, such as the feeling of turning while the aircraft is flying straight. You may respond to this sensation and apply control inputs for the perceived condition; thus, causing an undesired attitude. You may also lose track of which way is up unless a reliable visual reference is available.

Instrument flight is the skill used to overcome human limitations when you are unable to see the outside world. Despite what your senses tell you, when the turn needle of a turn co-ordinator or turn-and-bank indicator shows a turn in a certain direction, the aircraft is turning in that direction; when the nose of the miniature aircraft is below the horizon bar of the attitude indicator, the nose of the aircraft is below the real horizon. When the airspeed indicator shows a steady increase in speed during cruise flight, speed is increasing and the aircraft is most likely in a nose-down attitude. Therefore, during instrument flight you must have faith in the instrument indications and never react to an unconfirmed physical sensation, no matter how strong it is. The sooner you become comfortable with this, the more quickly the learning process develops.

As well, it is important to learn to relax while flying by instruments. First, hold the controls lightly as you cannot feel control pressure changes with a tight grip. Second, make smooth, small changes with a positive pressure. Third, with the aircraft properly trimmed, momentarily release all pressure on the controls when you become aware of tenseness. This will remind you that, if properly trimmed, the aircraft will remain in stable flight by itself.

It may also help if you visualize the attitude of the aircraft, as well as any movements taking place. If necessary, it may help to lower the pilot's seat for a better view of the instruments. If the aircraft is equipped with arm rests, use them. This allows more selective application of control pressures without having to constantly make allowances for the weight of your arm. The seat-belt should be fastened snugly; when the body is too free to move about in the seat, false sensory illusions become more acute and believable.

Aircraft Instruments

Throughout this chapter, we refer to three main groups of aircraft instruments. These are *control instruments, performance instruments*, and *navigation instruments*.

The Control and Performance Instruments

The two essential control instruments are the *attitude indicator* and the *tachometer* or *manifold pressure gauge*. The attitude indicator gives direct and immediate pitch and bank information. The tachometer (or manifold pressure gauge) gives direct power information. To set the attitude and power, refer directly to these instruments.

The performance instruments provide information that enables you to determine how the aircraft is behaving. For example, you may need to know if the altitude and heading are being correctly maintained. Two of the performance instruments, the altimeter and the heading indicator, can provide this information.

When the performance instruments show unwanted change in performance, your attention must be directed to the control instruments to allow you to make the required changes. For example, if the performance instruments indicate that the heading or the altitude has changed, attention is directed to the control instruments (attitude indicator and/or tachometer) while adjusting attitude and power sufficiently to make a correction. You then confirm that the required correction is in progress by referring again to the performance instruments (Fig. 2-68).

The performance instruments give both direct and indirect information as shown in the following table:

Instrument	Direct Information	Indirect Information
Airspeed Indicator	Airspeed	Pitch attitude
Altimeter	Altitude	Pitch attitude
Vertical Speed Indicator	Climb/descent rate	Pitch attitude
Heading Indicator	Heading	Bank attitude
Turn-and-Bank Indicator	Yaw, Co-ordination	Bank attitude
Turn Co-ordinator	Yaw, Roll, Co-ordination	Bank attitude
Magnetic Compass	Heading	Bank attitude

Fundamental Skills

Three fundamental skills are involved in all instrument flight manoeuvres: *instrument scan, instrument interpretation*, and *aircraft control*. A measure of your proficiency in instrument flying will be your ability to integrate these skills into unified, smooth, positive control responses to maintain a prescribed flight path.

Instrument Scan

Scanning, or cross-checking as it is sometimes known, is the continuous and logical observation of the flight instruments. A methodical and meaningful instrument scan is necessary to make appropriate changes in aircraft attitude and performance.

The Selective Radial Scan

To obtain information from the flight instruments, attention must be focused on each instrument long enough to read it. A split second is often long enough, and then the scan moves on to the next instrument. The attitude indicator is the central instrument, and it becomes the hub of the scan so that your attention usually returns to it after looking at a particular performance instrument. A diagram depicting the viewing pattern would show lines between the attitude indicator and the other instruments, just as spokes radiate from the hub of a wheel. This pattern of viewing the instruments is called a *radial scan*.

When using a radial scan, the instruments that provide the information most needed, given the task at hand, should be scanned more frequently. The other instruments are scanned less frequently for backup information. A radial scan that uses particular instruments to obtain the information necessary to carry out a particular task is called a *selective radial scan*. For example, when the task is to fly straight and level, the most important performance instruments are the heading indicator and the altimeter. Therefore, you would use the selective radial scan to focus on these two instruments more frequently. As shown in Fig. 2-69, the viewing pattern moves from the attitude indicator to the heading indicator, back to the attitude indicator, and from there to the altimeter and back to the attitude indicator.

Of the remaining performance instruments, the airspeed indicator and the vertical speed indicator give information to confirm the pitch attitude, and the turn co-ordinator and magnetic compass give information to confirm direction. These instruments are included in the scan but are scanned less frequently than the others.

Applying the Scanning Pattern

To develop the technique of always referring to the correct instrument at the appropriate time, you must continually ask yourself the questions:

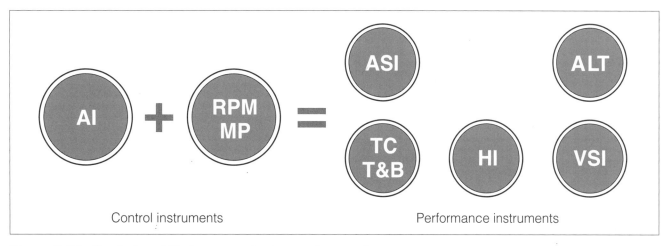

Figure 2-68   Control and Performance Instruments

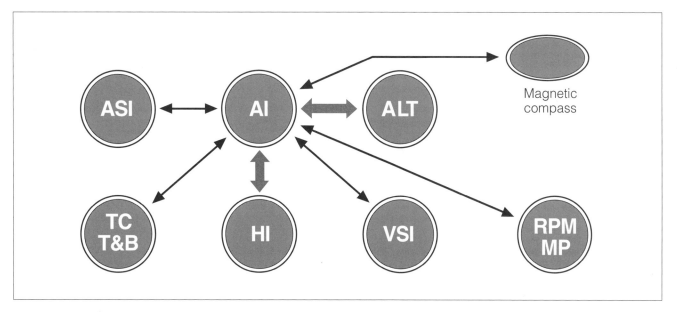

Figure 2-69   Straight-and-Level Flight

1. What information do I need?
2. Which instruments give me the needed information?
3. Is the information reliable?

Fig. 2-69, 2-70, 2-71 and 2-72 show examples of patterns that may be used for particular flight conditions.

Using Fig. 2-69, which depicts a selective radial scan pattern for straight-and-level flight, the answer to the question — What information do I need to fly straight and level? — is "heading and altitude." The answer to the next question — Which instruments give me the needed information? — is "the heading indicator and the altimeter." To answer the question — Is the information reliable? — you must first confirm the reliability of the heading indicator by referring to the turn co-ordinator and the magnetic compass, and second, confirm the reliability of the altimeter by referring to the vertical speed indicator and the airspeed indicator.

In all four diagrams, the coloured arrows depict the scanning pattern that you use most frequently to get the needed information. The black arrows indicate a less frequently repeated scan of the supporting instruments. The scan confirms that the main instrument indications are reliable, and determines if there is any trend toward an undesired flight condition. The supporting instruments that are more relevant to the task are viewed more frequently than the others. Using altitude control as an example, the vertical speed indicator would logically be viewed more frequently than the turn co-ordinator.

Fig. 2-70 shows the correct scan pattern for a straight climb. The coloured arrows show that most attention is given to heading and airspeed while climbing. The black arrows depict the less frequent scan used to confirm heading and attitude information.

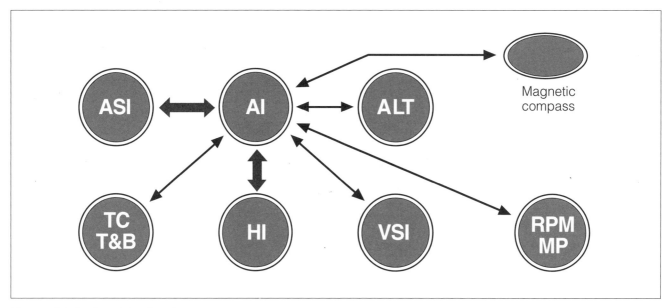

Figure 2-70   Straight Climb

In a straight climb, you can use the turn co-ordinator or turn-and-bank indicator and magnetic compass to confirm the heading. The vertical speed indicator and altimeter can confirm that the aircraft is performing as expected in the climb and can determine if the attitude indicator is reliable.

Fig. 2-71 shows the most appropriate scan as the aircraft nears the assigned altitude. At this point airspeed information becomes less important and altitude information becomes more important. During the transition from climbing to straight-and-level flight, use a scan pattern that gives more frequent attention to the attitude indicator with support from the heading indicator and altimeter.

As shown by the coloured arrows in Fig. 2-72, as the aircraft approaches the cruise attitude, airspeed information becomes more important. The airspeed indicator is scanned more often along with the attitude indicator. As the cruise attitude is established, allow the aircraft to accelerate to cruise speed before setting cruise power. Scan the heading indicator and altimeter often to establish and maintain straight-and-level flight.

Be sure to include all relevant instruments in your scan. There can be a tendency to fixate on one instrument when you become concerned about the information it gives you. For example, you might stare at an altimeter that reads 200 feet below assigned

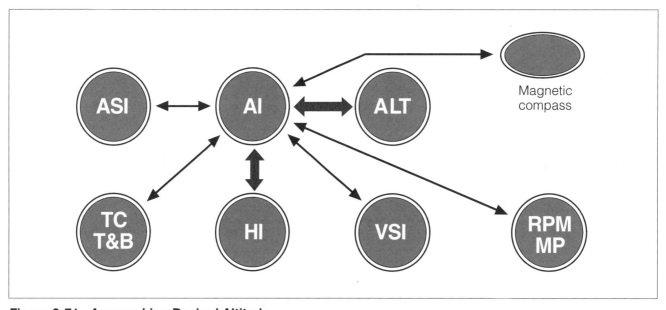

Figure 2-71   Approaching Desired Altitude

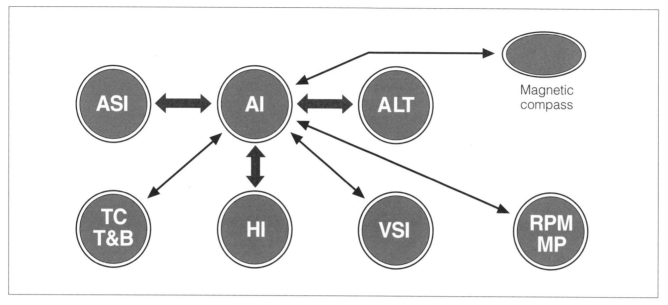

Figure 2-72   Level, Approaching Desired Airspeed

altitude and wonder how the needle got there. Meanwhile a heading change or other errors may occur.

Anticipate significant instrument indication changes following attitude changes. For example, when rolling out of a 180 degree turn using the attitude indicator, be sure to check the altimeter for altitude information.

Use all the instruments available for attitude information. You can maintain reasonably accurate altitude control with the attitude indicator, but the altitude cannot be held with precision without including the altimeter in the scan.

Instrument Interpretation

Instrument interpretation begins with the understanding of each instrument's operating principles. Then comes the application of this knowledge to the performance of the aircraft, and the flight condition in which it is operating. If the pitch attitude is to be determined, the attitude indicator, airspeed indicator, altimeter, and vertical speed indicator are used together to provide the necessary information. If the bank attitude is to be determined, the attitude, turn co-ordinator or turn-and-bank and heading indicator must each be interpreted.

For each manoeuvre, a combination of instruments must be interpreted to control aircraft attitude during the manoeuvre.

Attitude Plus Power Equals Performance

You must also interpret the instruments in relation to the performance capabilities of your aircraft. These capabilities can vary greatly from aircraft to aircraft as shown in Fig. 2-73. The combination of power and attitude is used in a light aircraft for a five-minute climb from near sea level. The attitude indicator shows the miniature aircraft two bar widths (twice the thickness of the miniature aircraft wings) above the horizon bar. With the power available in this particular aircraft and a selected attitude, the performance is shown on the instruments. That is, a climb of 500 feet per minute at 90 KT, and an altitude gain of 2,500 feet.

Now set up the identical picture in a jet aircraft. With the same aircraft attitude as in the first example, the vertical speed indicator in the jet reads 2,000 feet per minute, the airspeed indicates 300 KT and the altitude gain is 10,000 feet. As you learn the performance capabilities of an aircraft, you will interpret the instrument indications in terms of the attitude of the aircraft.

The phrase "Attitude plus power equals performance" summarizes the philosophy behind instrument flying. In other words, an aircraft's performance is the product of attitude and power. Performance is expressed in terms of airspeed, altitude, rate of climb or descent, or other criteria. If either attitude or power is changed, a change in performance will result.

Aircraft Control

With the instruments substituted for outside references, the necessary control responses and thought processes are the same as those for controlling the aircraft in visual flight. Control pressures should be smooth, making small corrections and waiting for the

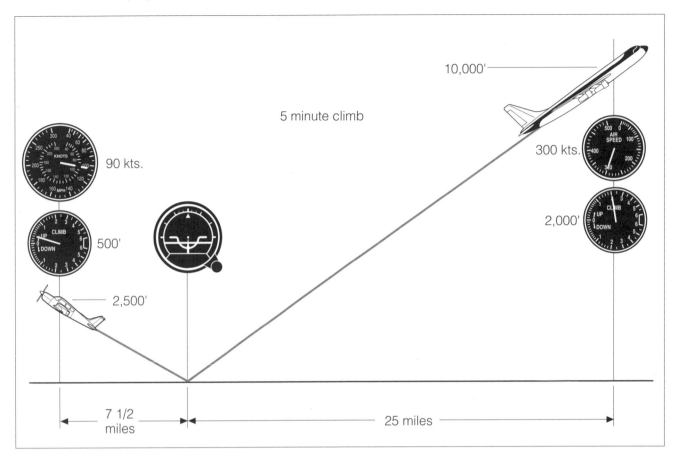

Figure 2-73   Attitude Plus Power Equals Performance

results. Anticipate the need to change power and attitude to arrive at desired airspeeds and altitudes when accelerating, decelerating, turning, climbing, descending, and levelling off.

The instrument scanning techniques described in the preceding examples are used in the following flight exercises, which are considered essential to controlling an aeroplane in instrument flight conditions.

Straight-and-Level Flight

Straight-and-level flight is accomplished by flying in a constant direction at a constant altitude. Although the most common application of straight-and-level flight is cruising, it can be achieved through a great range of pitch attitudes, power settings, and airspeeds.

Straight Flight

An aircraft is in straight flight when it is flying a constant heading. To maintain straight flight, the wings must be kept level with the horizon. If the wings are not level with the horizon, the aircraft will

turn, and this will be shown on the attitude indicator, heading indicator, and turn co-ordinator or turn-and-bank indicator. To level the wings, apply co-ordinated aileron and rudder control inputs.

The attitude indicator gives a direct indication of bank. On the standard attitude indicator, the angle of bank is shown pictorially by the relationship of the miniature aircraft to the attitude indicator bar, and by the alignment of the pointer with the banking scale at the top of the instrument. One advantage of the attitude indicator is that it offers at a glance an immediate indication of both pitch and bank attitude.

The performance instruments related to straight flight are the heading indicator and turn co-ordinator or turn-and-bank indicator. They display an indirect indication of bank because a banked aircraft has a natural tendency to turn. A rapid movement of the heading indicator in co-ordinated flight indicates a large angle of bank, whereas a slow movement reflects a small angle of bank, assuming the same airspeed in both instances.

The heading indicator gives accurate magnetic heading information only if it is set to correspond to the magnetic compass. An accurate reading can be taken from the basic magnetic compass only during unaccelerated flight. This means while flying straight-

and-level or during straight, stable climbs and descents. In turbulent air it may be necessary to take two or more readings from the compass and average the readings to determine the heading to set on the heading indicator.

The turn needle of the turn-and-bank indicator responds to yaw in such a way that a rapid yawing movement causes a large displacement of the turn needle, and a slow rate of yaw causes a small displacement. The turn co-ordinator responds to yaw in the same way, but it also responds to roll. You must keep this in mind when interpreting its indications. If the ball is centred and the needle of either of these instruments is deflected from the central position, you may logically conclude that the aircraft is banked in the direction of the needle deflection. Return to straight flight is accomplished through co-ordinated aileron and rudder pressures.

Abrupt use of aileron and rudder causes oscillation of the turn needle, making it difficult to interpret. When using the instrument to maintain straight flight, apply smooth control pressures. In turbulent air the turn needle oscillates from side to side. Therefore, you must average the fluctuations. When the deflection is greater on one side of centre than the other and the ball is centred, the aircraft is turning in the direction of the greater deflection. A turn will then be required to regain the desired heading.

Adverse yaw can be defined as *any yaw, regardless of origin, having an effect contrary to the interests of the pilot*. As the turn needle responds to yaw, preventing or controlling adverse yaw with rudder will bring the needle back to the desired position.

The ball of the turn-and-bank indicator or turn co-ordinator is actually a separate instrument, conveniently located under the turn needle because the two instruments are used together. It is affected by centrifugal force and gravity and indicates whether or not the aircraft is flying with some angle of side-slip. When these forces are balanced, the ball is centred within its glass tube and the manoeuvre being executed is said to be co-ordinated. However, if the ball is not in the centre, the aircraft is either slipping or skidding, and the side to which the ball has rolled indicates the direction of the slip or skid. To differentiate between a slip and a skid consider the following:

Slip.   If the needle is centred and the ball is displaced in either direction from centre, the aircraft is slipping. Fig. 2-74 left shows a slip to the right.

If both the needle and ball are displaced to the same side of centre, the aircraft is in a slipping turn. Fig. 2-74 right shows a slipping turn to the left.

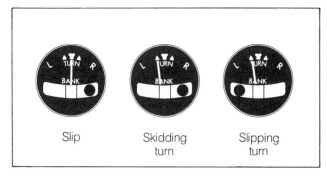

Figure 2-74   Slipping and Skidding

Skid.   If the ball is displaced to one side and the needle is displaced to the opposite side, the aircraft is in a skidding turn. Fig. 2-74 centre shows a skidding turn to the left.

When the needle of a turn-and-bank indicator or turn co-ordinator is displaced from centre, the aircraft is yawing in the direction indicated. When the ball is kept centred, the needle shows the direction of bank and the direction of turn. However, if the ball is left or right of centre, the direction the needle is deflected from centre may not necessarily be the direction in which the aircraft is banked.

Under most instrument flight conditions the ball should be centred. If the ball is displaced to the left it may be centred by right aileron pressure, but this may introduce an undesirable bank angle or turn. The ball may also be centred by left rudder pressure, but this too may introduce unwanted yaw and subsequent turning. The correct response is to co-ordinate the application of rudder and aileron to produce the desired flight path with the ball centred.

Level Flight

An aircraft is in level flight when it is flying at a constant altitude. At a constant cruise power setting, a deviation from level flight will result if the nose is pitched up or down. The instruments that will respond to this are the attitude indicator, altimeter, vertical speed indicator, and airspeed indicator. The attitude indicator gives a direct indication of pitch attitude. A desired pitch attitude is attained using the elevator control to raise or lower the miniature aircraft in relation to the horizon bar. This corresponds to the way the pitch attitude is adjusted in visual flight by raising or lowering the nose of the aircraft in relation to the natural horizon.

If the altimeter and vertical speed indicator show a climb or descent, the aircraft is not in level flight, and an attitude correction is necessary to maintain altitude at the selected power setting. When a pitch error is detected, corrective action should be taken promptly but with light control pressures and with three distinct changes of attitude. First, a change of attitude to stop the needle movement; second, a change of attitude to return to the desired altitude; and third, a change of attitude to stop the correction and assume the desired flight attitude. Use small, smooth attitude changes to correct for small rates of climb or descent, and retrim. At this point, it may be desirable to reset the position of the small aeroplane on the attitude indicator so that it is superimposed on the horizon line.

When the altimeter and vertical speed indicator show that the aircraft is not in level flight, it is important to note the rate at which the aircraft is climbing or descending and apply control inputs accordingly. If the altitude is changing slowly, it indicates that the attitude is probably close to that required. Small control inputs will likely be sufficient to correct. If a rapid rate of climb or descent is noted, it indicates that a large deviation in attitude has taken place. Larger control inputs will be needed to correct this condition, and a change in power may be required.

The vertical speed indicator gives an indirect indication of pitch attitude and is both a trend and a rate instrument. As a trend instrument, it shows the initial vertical movement of the aircraft which, disregarding turbulence, can be considered a reflection of pitch change. To maintain level flight, use the vertical speed indicator in conjunction with the altimeter and attitude indicator. Note any up or down trend of the needle from zero and apply a very light corrective elevator pressure. If control pressures have been smooth and light, the needle will react promptly and slowly.

Used as a rate instrument, the vertical speed indicator's *lag* characteristics must be considered. Lag refers to the time delay before the needle attains a stable indication following a pitch change. Lag is directly proportional to the speed and magnitude of a pitch change. At a constant power setting, if a slow, smooth pitch change is initiated, the needle will move with minimum lag and then stabilize when the rate of climb or descent is steady. A large and abrupt pitch change will produce erratic needle movement and also introduce greater time delay before the needle stabilizes. Take care not to "chase the needle" when flight through turbulent conditions produces erratic needle movements.

When correcting for an altitude error, keep in mind that the amount of the error governs the rate at which you should return to the required altitude. A rule of thumb is to make an attitude change that will result in a vertical speed that is approximately double the error in altitude. For example, if the altitude error is 100 feet, the rate of return should be approximately 200 feet per minute. If a large altitude error is noted, the rate of return should be correspondingly greater,

but the airspeed should not exceed the speed for optimum rate of climb or descent for the aircraft.

The airspeed indicator gives an indirect indication of pitch attitude. At a constant power setting and pitch attitude, the airspeed remains constant. If the nose of the aircraft lowers, the airspeed increases and the nose should be raised. If the nose rises, airspeed decreases and the nose should be lowered. A rapid change in airspeed indicates a large pitch change, and a slow change of airspeed indicates a small pitch change. The apparent lag in airspeed indications with pitch changes varies among different aircraft and is due to the time required for the aircraft to accelerate or decelerate when the pitch attitude is changed. There is no appreciable lag due to the construction or operation of the instrument.

To change airspeed during level flight, select a power setting appropriate to the desired speed and adjust the attitude to maintain level flight (Fig. 2-75). Once again, trim the aircraft so that it is stable, and reset the attitude indicator.

When making pitch attitude corrections in response to information taken from the altimeter, vertical speed indicator, or airspeed indicator, attention should be directed to the attitude indicator while the control inputs are being made. When raising or lowering the nose, the amount of attitude adjustment is monitored by watching the response of the attitude indicator. When a new attitude has been attained, a cross-check of the performance instruments will show if it is the correct attitude to maintain level flight.

Power Control

While in straight-and-level flight, any change in power setting results in a change in airspeed or altitude. When the airspeed is constant, an increase in power will cause the aircraft to climb, and a decrease in power will cause the aircraft to descend. If the altitude

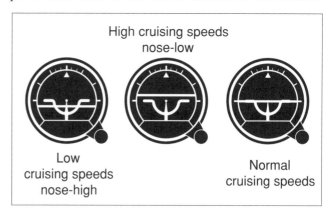

Figure 2-75 Attitudes at Different Cruising Speeds

is maintained, a change in power will affect the airspeed.

To increase airspeed in straight-and-level flight, power is added: to decrease airspeed, power is reduced. To maintain altitude at an increased power setting, the nose must be lowered. When power is decreased, the nose must be raised. In addition, changes in power tend to cause adverse yaw, which must be controlled with rudder.

Power control and airspeed changes are much easier when you know the approximate power settings necessary to maintain various airspeeds in straight-and-level flight. A rule of thumb for airspeed control is 100 RPM, or one inch of manifold pressure, produces approximately a 5 KT change in airspeed. For example, consider an aircraft that requires 2,300 RPM to maintain 120 KT in straight-and-level flight. If the airspeed is to be reduced to 100 KT, you should reduce power by 400 RPM to 1,900 RPM. As this is only an approximate rule, a second minor adjustment in power may be necessary.

Consider the following points:

Heading

1. Be sure to scan the heading indicator, especially during changes in power and pitch attitude.
2. Interpret heading changes correctly to avoid making heading changes in the wrong direction.
3. Correct for *all* heading deviations.
4. Use small bank angles for small heading corrections.

Pitch

1. Check the attitude indicator and make any necessary adjustments to the miniature aircraft for level flight indication at normal cruising airspeed.
2. Maintain a continuous scan and accurately interpret all the pitch instruments.
3. Make pitch corrections as soon as an altitude deviation is noticed.
4. As soon as a pitch correction is required, attention should be focused on the attitude indicator, while the control inputs are being applied.

Power

1. Apply the power settings appropriate to various airspeeds or drag configurations.
2. Make smooth throttle movements.
3. Lead with power when making airspeed changes. For example, during an airspeed reduction in level flight, adjust the throttle to maintain the slower speed before the airspeed reaches the desired level.

4. Maintain a continuous scan. This will result in smooth airspeed changes.

Trim

1. Use trim only to relieve pressure on the controls.
2. Use trim frequently if required, and in small amounts.

Climbing

The ability to climb at a particular airspeed or a given rate of climb, or both, is essential to obtain the best climb performance from the aircraft. To accomplish this, it is a good practice to estimate the airspeed and power setting appropriate for the desired performance. The rules of thumb given at the end of this section will be useful in making these estimates.

Entry

To enter a constant airspeed climb from cruising flight, raise the nose of the miniature aircraft to the approximate nose-up indication for the predetermined climbing speed. Control pressures will vary as the aircraft decelerates. Advance the power to the climb power setting after the nose-up attitude is established and the airspeed approaches climbing speed. If the transition from level flight is smooth, the vertical speed indicator will show an immediate trend upward and stop at a rate of climb appropriate to the stabilized airspeed and attitude. Trim as necessary.

Scan the attitude indicator and airspeed indicator to ensure that the desired attitude is maintained. If the climb attitude is correct, the airspeed will stabilize at the desired speed. If the airspeed is low or high, make an appropriate pitch correction. Scan the heading indicator to ensure that the desired heading is maintained. Be prepared for the yawing tendency that normally occurs during a climb and keep straight with rudder.

To enter a climb at a predetermined rate of climb and speed, estimate the attitude and power setting to achieve the desired performance. Set the climb attitude using the attitude indicator. Adjust the power to the predetermined setting for the climb. Scan the attitude indicator and the airspeed indicator to ensure that the desired attitude is maintained and adjust the attitude as necessary to maintain the predetermined airspeed. Cross-check the vertical speed indicator with the attitude indicator to ensure that the desired rate of climb is maintained. Adjust the power as necessary to increase or decrease the rate of climb. A change of

100 RPM, or one inch of manifold pressure changes airspeed approximately 5 KT or the rate of climb by approximately 100 feet per minute. Scan the heading indicator to ensure that the desired heading is maintained.

Levelling Off

Levelling off from a climb must be started before reaching the desired altitude. The amount of lead varies with rate of climb. If the aircraft is climbing it will continue to climb at a decreasing rate throughout the transition to level flight. An effective practice is to lead the altitude by 10 percent of the vertical speed shown.

To level off at cruising airspeed, apply smooth, steady forward elevator pressure to lower the nose to the cruise attitude on the attitude indicator. As the nose is lowered to maintain altitude, the vertical speed gradually decreases toward zero and the airspeed increases. Continuing forward pressure in pitch control is needed as the airspeed increases. When the airspeed reaches cruising speed, set cruise power and trim. Cross-check the altimeter and heading indicator to confirm that the required altitude and heading are maintained.

Descending

While descending, the objective may be to maintain a constant airspeed, a constant rate of descent, or both.

Entry

To enter a constant airspeed descent from cruising flight, reduce the power as required for the descent and maintain the cruise attitude. As the airspeed decreases to the desired airspeed, adjust the pitch attitude to maintain that airspeed and trim as necessary.

Scan the attitude indicator and airspeed indicator to ensure that the desired attitude is maintained. If the attitude is correct, the airspeed will stabilize at the desired speed. If the airspeed is low or high, make a small pitch correction. Scan the heading indicator to ensure that the desired heading is maintained.

To enter a descent at a constant rate, estimate the attitude and power setting for the descent. If the descent will be carried out at a reduced speed, reduce the power, maintain the cruise attitude and allow the airspeed to decrease as necessary before adjusting the attitude for the descent. If cruise speed is to be maintained during the descent, reduce power as required,

lower the nose to maintain cruise airspeed and control yaw. Scan the attitude indicator and the airspeed indicator to ensure that the desired attitude is maintained and the vertical speed indicator to ensure that the desired rate of descent is maintained. If necessary, adjust the attitude to maintain the predetermined airspeed and adjust the power to increase or decrease the rate of descent. A change of 100 RPM, or one inch of manifold pressure, changes airspeed approximately 5 KT or the rate of descent by approximately 100 feet per minute. Scan the heading indicator to ensure that the desired heading is maintained.

Levelling Off

Levelling off from a descent must be started before reaching the desired altitude. The aircraft will continue to descend at a decreasing rate throughout the transition to level flight. An effective practice is to lead the altitude by 10 percent of the vertical speed.

To level off from an 800 foot per minute descent for example, lead the altitude by approximately 80 feet, adjust the attitude and simultaneously increase the power to the required setting. The vertical speed will decrease toward zero. Scan the attitude indicator, altimeter, and heading indicator to confirm that the desired heading and altitude are maintained. Make small adjustments to the attitude as necessary to maintain altitude, and trim.

Turns

When flying with reference to instruments, control inputs, turn entry and recovery procedures, and use of power are the same as when flying with outside visual references. When using instrument reference the typical turn is accomplished using an angle of bank that results in a turn rate of 3 degrees per second. This is referred to as a rate one or standard rate turn. Normally the angle of bank used during instrument reference turns does not exceed 30 degrees.

To produce a rate one turn, first estimate the angle of bank required using the formula in the suggested rules of thumb on page 158. Enter the turn referring to the attitude indicator miniature aircraft and bank scale. Use co-ordinated aileron and rudder inputs to roll to the desired angle of bank. Because the angle of bank calculated from the formula is approximate, ensure the turn needle on the turn co-ordinator or turn-and-bank indicator indicates a rate one turn. Note and maintain the angle of bank shown on the banking scale of the attitude indicator when the turn needle is directly under the standard rate index. Fig. 2-76 shows a turn-and-bank indicator and turn co-ordinator indicating rate one turns.

At this point a small amount of nose-up pitch is usually required to maintain altitude. Refer to the dot on the attitude indicator and raise it slightly with reference to the horizon bar by applying elevator. After the turn is established, fluctuations of the turn needle must be controlled with co-ordinated rudder and aileron inputs. During the turn, cross-check the attitude indicator with the altimeter and the heading indicator, with occasional reference to the turn needle.

To return to straight flight, lead the desired heading as shown in the rules of thumb. Refer to the attitude indicator while rolling the wings level, and use the same rate of roll for the recovery as for the entry. As the wings roll to the level position, the aircraft may show a tendency to climb if the back pressure on the elevators is not released. At this point, cross-check the altimeter frequently and adjust the pitch attitude to maintain altitude.

Figure 2-77   Steep Left Turn

instead of raising the nose. If the vertical speed indicator and the altimeter indicate a descent and the airspeed is increasing despite increased backward pressure on the control column, reduce the bank angle to one where the elevator control can be used to raise the nose and restore the aircraft to level flight. Then re-establish the required bank angle.

During recovery from steep turns, lead the desired heading as shown in the rules of thumb and refer to the attitude indicator while rolling the wings level. As the wings roll to the level position, pitch the nose down slightly to maintain altitude and adjust the power as required.

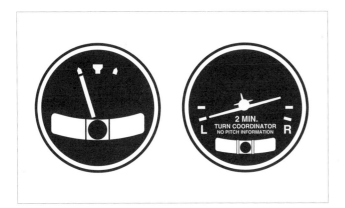

Figure 2-76   Rate One Turns

Steep Turns

Enter a steep turn using the same control inputs as for any other turn, and be prepared to scan rapidly for pitch and bank attitude information as the turn steepens. As the angle of bank exceeds 30 degrees, pitch the nose slightly above the horizon bar on the attitude indicator to maintain altitude and increase power as required to maintain a selected airspeed (Fig. 2-77). The power necessary to maintain a desired altitude and airspeed increases as the bank increases. When the required angle of bank is reached, keep it constant with aileron and control yaw with rudder.

As the bank angle increases, pitch corrections necessary to maintain altitude will require increasingly stronger elevator pressure. If bank angle continues to increase, a point will be reached where further application of back elevator pressure tightens the turn

Change of Airspeed in Turns

Changing airspeed in turns is an effective manoeuvre for increasing proficiency in all basic instrument skills as it involves simultaneous changes in all components of control. Proper execution requires rapid scanning and interpretation as well as smooth control. Proficiency in the manoeuvre will also contribute to confidence in the instruments during attitude and power changes involved in more complex manoeuvres. Pitch and power control techniques are the same as those used during changes in airspeed in straight-and-level flight.

The angle of bank necessary for a given rate of turn is proportional to the true airspeed. The angle of bank must be varied in direct proportion to the airspeed if a constant rate of turn is to be maintained. For example, during a reduction of airspeed, decrease the angle of bank. The pitch attitude must also be increased to maintain a level turn. The altimeter and turn needle indications should remain constant throughout the turn.

While reducing airspeed in a turn, the rate of scanning must be increased as you reduce power. As the aircraft decelerates, check the altimeter and vertical speed indicator for pitch changes and the bank instruments for bank changes. Adjust pitch attitude to maintain altitude, and as you approach the required airspeed adjust the power setting to maintain it. Trim is important throughout the manoeuvre to relieve control pressures. Frequent cross-check of the attitude indicator is essential to keep from overcontrolling and to provide approximate bank angles appropriate to the changing airspeeds.

Suggested rules of thumb:

1. The approximate angle of bank to produce a rate one turn may be calculated by using the following formula: (IAS in KT divided by 10) + 7 = bank angle. Add 5 instead of 7 for statute miles per hour.
2. Use small angles of bank to make small heading changes. Usually a bank angle equal to half the number of degrees of heading change will suffice. In any case, limit bank angle to no greater than that required for a rate one turn.
3. To roll out of a turn on a selected heading, lead the heading by half the angle of bank, e.g., if using a 30 degree bank, begin the roll-out 15 degrees before reaching the desired heading.

Partial Panel

The term *partial panel* refers to instrument flying while the attitude indicator and heading indicator are either not fitted on the panel, or are unserviceable. With a full panel, the attitude of the aircraft is determined either by reading direct information from the attitude indicator or by interpreting indirect information from the performance instruments. While flying an aircraft without an attitude indicator, you must determine the pitch attitude by interpreting airspeed, altitude, and vertical speed indications. For example, in straight-and-level flight with cruise power set, the airspeed can be expected to be the same as it has always been for that condition. By establishing an attitude and power setting that results in the expected cruise speed, the aircraft will be in, or very close to, the cruise attitude. The altimeter and vertical speed indicator will then confirm if the attitude is correct for level flight.

Without an attitude or heading indicator an important source of direction and bank information is lost. However, you know that when the ball of the turn-and-bank indicator or turn co-ordinator is centred and the aircraft is on a constant heading the wings of the aircraft will be level. The magnetic compass is the performance instrument used to confirm that a constant, correct heading is being maintained. Keep straight by using co-ordinated aileron and rudder inputs to keep the ball centred and eliminate any yaw indications shown on the turn needle.

Scanning the Partial Panel

When flying without the attitude and heading indicators, as with full panel, you must scan the instruments to ensure that the aircraft is being flown as required. However, the scan must be somewhat modified to use other instruments to determine the attitude information needed to control the aircraft. The principle of instrument flying remains unchanged. You continue to control the aircraft in accordance with the formula "attitude plus power equals performance." As with full panel, you must continually answer the basic questions:

1. What information do I need?
2. Which instruments give me the needed information?
3. Is the information reliable?

Straight-and-Level Flight

To fly straight in an aircraft with no heading indicator, frequently refer to the turn needle to detect any heading change (Fig. 2-78). Controlling an aircraft in straight flight by means of the turn co-ordinator or turn-and-bank indicator requires a return to basic control principles, i.e., control yaw with the rudder and keep the wings level with ailerons. Therefore, when flying straight using the turn co-ordinator or turn-and-bank indicator, prevent yaw with appropriate rudder pressure, and keep the wings level with appropriate aileron pressure. The needle will remain centred while heading is maintained as no turn exists. If the ball is kept centred the needle will always indicate two important aspects of flight: the direction in which the aircraft is turning and the rate at which it is turning. This information is cross-checked with the magnetic compass to verify or determine whether the heading is being maintained.

To fly level using partial panel, frequently scan the altimeter and vertical speed indicator for early detection of altitude errors and occasionally scan the airspeed indicator for feedback on the pitch attitude (Fig. 2-78).

In turbulent conditions you may have to average the movement of the turn needle or magnetic compass.

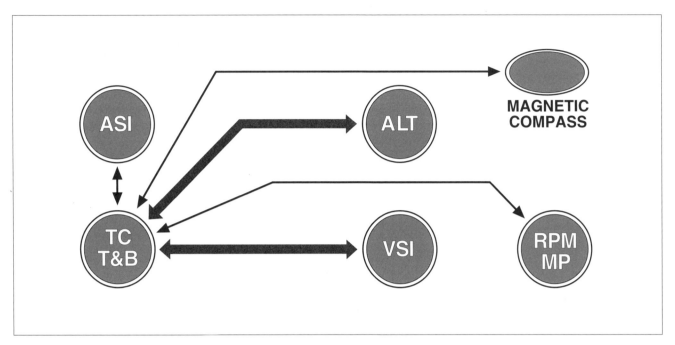

Figure 2-78   Straight-and-Level Flight

Climbing

To enter a climb using partial panel, use normal, smooth control pressures and observe the indications of the airspeed indicator and the vertical speed indicator as a nose-up attitude is being established. They may take a few moments to react. When they begin to register the changes associated with climb entry, hold the attitude constant, apply climb power, and trim. When the airspeed is stable, make minor pitch adjustments as required.

During the climb, the airspeed indicator must be interpreted for pitch information. Use small pitch adjustments for small airspeed corrections, anticipate yaw, and keep straight with rudder. Check the magnetic compass occasionally to ensure the desired heading is maintained. The scan for a straight climb includes frequent reference to the airspeed indicator and turn co-ordinator or turn-and-bank indicator, and less frequent reference to the altimeter, vertical speed indicator, and the magnetic compass (Fig. 2-79).

To level off, the altimeter should be scanned more frequently as the aircraft approaches the required altitude (Fig. 2-80). Apply the rule of thumb for leading the altitude. Sufficient forward elevator control should be applied to stop the altimeter movement at the desired altitude. Cross-check the vertical speed indicator. As shown by the coloured arrows in Fig. 2-81, as the aircraft approaches the required airspeed the airspeed indicator is scanned more frequently. When the airspeed increases to cruise speed, set cruise power and trim.

Descending

To enter a descent using partial panel, adjust the power and attitude as you would when using full panel. After the power is set and the airspeed approaches descending speed, lower the nose to maintain the desired airspeed. The airspeed indicator is interpreted to give pitch information for the descent using the same technique as for the climb. Use small pitch adjustments for small airspeed corrections, anticipate yaw, and keep straight with rudder.

The scan for the straight descent is the same as the scan for the straight climb (Fig. 2-79).

To level off, apply the correct lead and adjust the pitch attitude to stop the altimeter at the desired altitude. Set cruise power without delay to avoid any decrease in airspeed and trim.

Turns

While flying on partial panel, most turns are made at a rate not exceeding rate one (3 degrees per second). Without an attitude indicator, refer to the turn needle of the turn-and-bank indicator or turn co-ordinator while rolling the aircraft into the turn. Maintain gentle aileron control pressure until the turn needle reaches the rate one index, then neutralize the ailerons. The scan for the turn includes frequent reference to the turn needle for bank information and the altimeter and airspeed indicator for pitch information (Fig. 2-82). To return to straight flight, roll the wings level using the same rate of roll as for the entry.

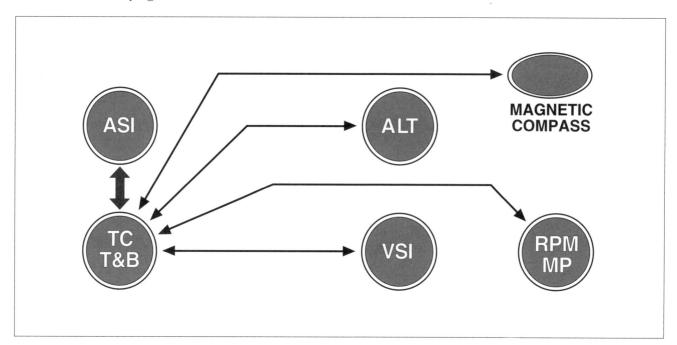

Figure 2-79 Straight Climb and Descent

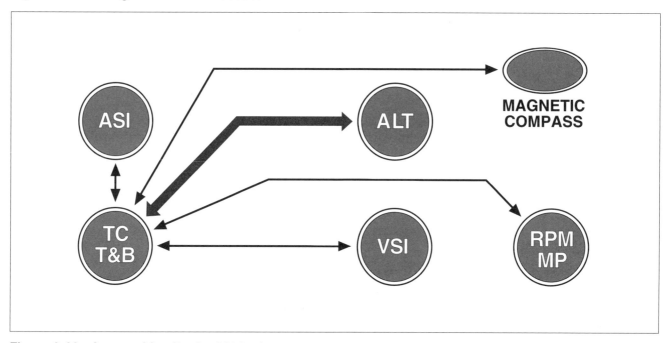

Figure 2-80 Approaching Desired Altitude

The magnetic compass is not reliable for establishing a heading during a turn. A more reliable turn to a heading can be made by turning at a known rate of turn for a known period of time. This is called a timed turn and, as such, requires that the clock be included in your scan. For example, a turn through 90 degrees at rate one takes 30 seconds. Begin the roll-in to establish a rate one turn when the clock second hand passes a prominent point. At the last second of the timing, initiate the roll-out to wings level at the same rate as you rolled in. After the aircraft is established in straight flight, check the magnetic compass and make any required small heading corrections with a small rate of turn.

Errors in the turn-and-bank indicator or turn co-ordinator needle indications may exist due to insufficient or excessive rotor speed or inaccurate adjustment of the calibrating spring. The accuracy of the rate one indication on these instruments may be determined by timing the heading change in a turn while holding the needle on the rate one mark. For example, a 30 degree heading change should take 10

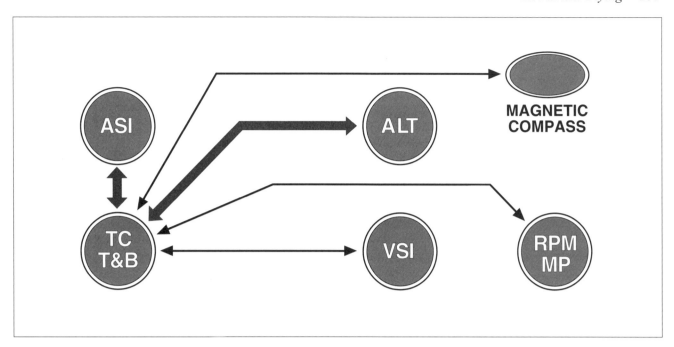

Figure 2-81   Level, Approaching Desired Airspeed

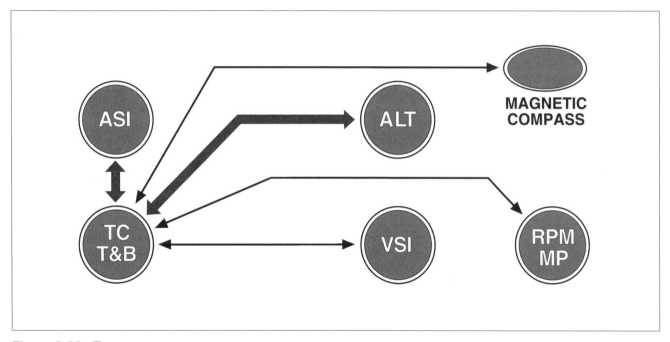

Figure 2-82   Turns

seconds. Any errors can be eliminated in future turns
by holding slightly more or less bank as required.

Unusual Attitudes and Recoveries

As a general rule, whenever you note an instrument
rate of movement or indication other than those you
associate with the basic instrument flight manoeuvres

already learned, assume an unusual attitude and
increase the speed of the cross check to confirm the
attitude.

Unusual attitudes can result from a number of conditions, such as turbulence, disorientation, preoccupation with cockpit duties, incorrect scan techniques,
errors in instrument interpretation, and instrument
malfunctions. As unusual attitudes are not performed
intentionally, except in training, they happen unexpectedly. When an unusual attitude is noted on the
cross-check, the immediate problem is not how the

aircraft got there, but what it is doing and how to get it back to straight-and-level flight as quickly as possible. As was pointed out in the introduction to this chapter, you can't rely on what your senses tell you. Recognition and recovery must be carried out using the following procedures.

Recognition

Nose-up attitudes (Fig. 2-83) are identified by the rate and direction of movement of the altimeter, vertical speed, and airspeed needles. Nose-down attitudes are shown by the same instruments, but in the opposite direction. If the ball of the turn co-ordinator or turn-and-bank indicator is centred, the direction of turn indicated by the needle will also be the direction in which the aircraft is banked.

Recovery

Recovery should then be made by reference to the airspeed indicator, turn co-ordinator or turn-and-bank indicator, altimeter and vertical speed indicator.

In moderate unusual attitudes it may be possible to establish level flight using the attitude indicator. However, in extreme attitudes the information shown on the attitude indicator may become unreliable as a reference for recovery. The following rules emphasize interpretation of the attitude from the performance instruments.

Procedure for Recovery from Unusual Attitudes

Nose Down:

1. Reduce power to prevent excessive airspeed and loss of altitude.
2. Level the wings by applying co-ordinated aileron and rudder pressures to centre the turn needle and ball.
3. Apply smooth back elevator pressure to return to level flight.
4. When the airspeed stops increasing, you are at or near level flight; stop the back elevator pressure.

Nose-Up:

1. Increase power to prevent further loss of airspeed.
2. Simultaneously apply forward elevator pressure to lower the nose to prevent a stall.

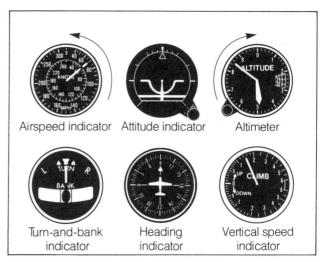

Figure 2-83   Nose-Up Unusual Attitudes

3. Level the wings by applying co-ordinated aileron and rudder pressure to centre the turn needle and ball.
4. When the airspeed stops decreasing, you are at or near level flight; stop the forward elevator pressure.

As the initial control inputs may have to be large, continue with a fast cross-check to detect possible overcontrolling. When the movement of the altimeter and airspeed indicator needles stop, the attitude is approaching the cruise attitude. Should the needles stop and then rotate in the opposite direction, the aircraft has passed through the cruise attitude. As the indications of the airspeed indicator, altimeter and turn-and-bank indicator stabilize, check the attitude indicator for proper functioning before incorporating it into the scan. A malfunction in this instrument may have been the cause of the unusual attitude. The attitude indicator, turn needle, and ball should be checked to determine bank attitude. Corrective control pressures should be applied as needed. The ball should be centred, as skidding and slipping sensations can easily aggravate disorientation and delay recovery.

After recovery from an unusual attitude, it is important to continue with a rapid cross-check of instruments to prevent entry into a secondary unusual attitude. In addition, it may be necessary to climb or descend to return to a safe altitude as soon as the aircraft is under control.

Spin Recovery

The spin is the most critical unusual attitude of all, not necessarily because of the manoeuvre itself but because of the disorientation that usually accompanies it.

The first requirement for spin recovery is to determine the direction in which the aircraft is spinning. The only reliable instrument for this purpose is the turn needle of the turn-and-bank indicator or the turn co-ordinator. The needle will show a deflection in the direction of the spin, and the altimeter will show a rapid loss of height. To differentiate between a spin and a spiral, check the airspeed. It will be low (near the stalling speed) in a spin and high and increasing in a spiral.

Having determined the direction of the spin, disregard the ball instrument temporarily and use the recovery techniques specified in the Aircraft Flight Manual. The procedure outlined below is suitable for most small aeroplanes and may be used in the absence of a procedure recommended by the manufacturer.

1. Power to idle, neutralize ailerons.
2. Apply and hold full rudder opposite to the direction of rotation, e.g., if the turn needle is full right apply full left rudder.
3. Just after the rudder reaches the stop, move the control column positively forward far enough to break the stall. Full-down elevator might be required in some aeroplanes.
4. Hold these control inputs until the turn needle starts moving back to the centre, indicating that the spin has stopped.
5. Neutralize the rudder (needle at or near centre).
6. Apply back pressure on the control column to ease the aircraft out of the dive.
7. When the airspeed begins to decrease, hold the pitch attitude constant and apply power to resume cruising flight. Keep the turn needle and ball centred with co-ordinated control pressures.

Basic Instrument Flight Patterns

A good way to continue practice once you have achieved a degree of proficiency on instruments is to simulate basic flight patterns used in Instrument Flight Rules (IFR) procedures. Knowing the flight requirements of these procedures will prove advantageous if you intend to extend your instrument flight training beyond an elementary stage. The actual procedures presume a starting point or fix such as a Non-Directional Beacon (NDB) or a Very High Frequency Omni Range (VOR). For practice purposes a visual geographic fix may also be used.

At this stage timing, precision turns, and the maintenance of specific airspeeds and altitudes on instruments are the primary objectives. The holding pattern outlined in Fig. 2-84 demands accurate timing and

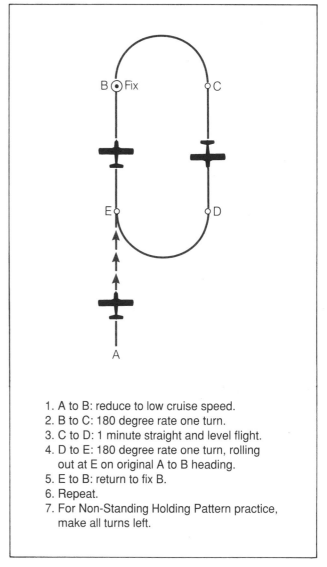

1. A to B: reduce to low cruise speed.
2. B to C: 180 degree rate one turn.
3. C to D: 1 minute straight and level flight.
4. D to E: 180 degree rate one turn, rolling out at E on original A to B heading.
5. E to B: return to fix B.
6. Repeat.
7. For Non-Standing Holding Pattern practice, make all turns left.

Figure 2-84   Standard Holding Pattern (all turns right)

instrument flying skill to stay within the bounds of the pattern at the assigned altitude.

The procedure turn shown in Fig. 2-85 is typical of those used in standard instrument approach procedures. For non-standard procedures, make a right turn at B and left turns at all others.

Radio Navigation

Very High Frequency Omni-directional Range

One common means of radio navigation is the Very High Frequency Omni-directional Range (VOR). This system consists of many strategically placed ground

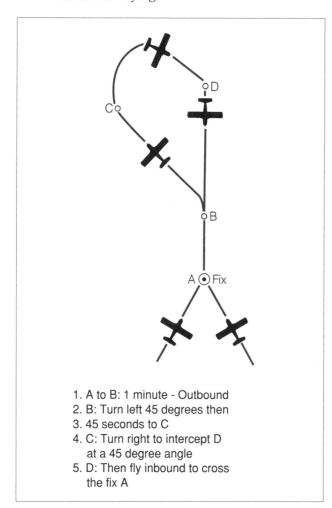

1. A to B: 1 minute - Outbound
2. B: Turn left 45 degrees then
3. 45 seconds to C
4. C: Turn right to intercept D
   at a 45 degree angle
5. D: Then fly inbound to cross
   the fix A

Figure 2-85   Procedure Turn

stations that transmit navigational signals in the Very High Frequency (VHF) range 108.10 to 117.95 MHz. An airborne VOR receiver converts these signals into visual indications which may be used for accurate navigation. A typical airborne VOR installation is shown in Fig. 2-88.

Like all VHF signals, VOR signals are subject to line-of-sight transmission, hence the greater the height the greater the range of the signal. For this reason, caution must be exercised when flying close to the ground or within mountainous regions. Line-of-sight range at 1,000 feet AGL is approximately 39 miles; at 2,000 feet AGL, 54 miles; at 3,000 feet AGL, 66 miles; at 4,000 feet AGL 77 miles.

Before using the VOR system, tune in the desired frequency and listen to the identification signal to be sure the right station has been selected. VOR stations are identified by a three letter morse code or a voice identification.

Most large airports have a VHF omni-test frequency (VOT), which enables you to determine the serviceability and accuracy of the receiver. To test the receiver,

select the VOT frequency. You should hear a rapid succession of morse code dots or the local Automatic Terminal Information Service (ATIS) broadcast (see Chapter 30). With the omni bearing selector (OBS) set at 360 degrees you should get a FROM indication in the sense or TO-FROM indicator and the course deviation indicator (CDI) should be centred. With the OBS set at 180 degrees you should get a TO indication in the sense indicator and the CDI should be centred. In either case if the OBS indicator is reading within plus or minus four degrees the system is acceptable for Instrument Flight Rules (IFR) flight. Finally, turn the OBS 10 degrees either side of needle centre and full deflection of the CDI should result.

Although there are no error limits specified that preclude use in VFR flight, you should be cautious and not rely on the VOR if you suspect inaccuracies. However, any attempt to apply an error found during the VOR check will complicate VOR navigation as there is no guarantee the error is the same throughout 360 degrees. In any event the OBS setting must be the same as your required track.

VOR provides you with 360 different magnetic tracks (one for each degree in a circle) from which you can select the track that best suits your needs to fly TO or FROM a VOR station. These tracks are called *radials*.

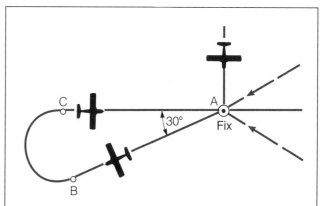

1. Approach fix A in level flight
2. A: Take up a heading 30 degrees "off" the outbound track; maintain this heading for one minute at low cruise speed to B
3. B: Turn right to roll out on the inbound track at C
4. C: Descend 800 feet and level off
5. A: After passing fix, descend 500 feet at approach speed.

Note: Turns and timing must be adjusted to compensate for wind effect.

Figure 2-86   Tear-Drop Procedure Turn

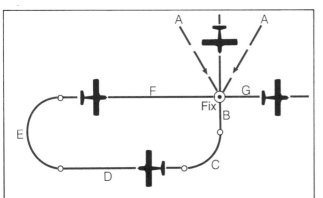

A: Approach fix in level flight; cross fix,
B: Fly 90 degrees to outbound heading for 30 seconds at low cruise speed,
C: Rate one turn; roll out on outbound heading,
D: Hold outbound heading for 1½ minutes, (reciprocal of inbound heading),
E: Rate one turn; roll out on published inbound heading and descend 700 feet as soon as aircraft is on a definite inbound heading, and level off,
F: Inbound heading to fix; after crossing fix inbound, descend 500 feet (G) at approach speed and level off.

Note: Adjust all turns and timing to compensate for wind effect.

Figure 2-87   Direct Entry Procedure

Examination of a navigation chart will show the VOR stations surrounded by a compass rose that enables you to identify the radial or desired track in degrees magnetic. All radials are identified as bearings from the station, for example, the 090 degree radial extends east of the station. The TO-FROM indicator is used to indicate whether an aircraft is on a bearing toward or from the VOR station. If the TO-FROM indicator reads TO with the CDI centred, the heading shown on the OBS represents the track from the aircraft TO the station. In Fig. 2-89 the aircraft is on the 090 degree radial flying a magnetic track of 270 degrees TO the VOR. If the TO-FROM indicator reads FROM the heading shown on the OBS represents the track of the aircraft FROM the station. At every position, two VOR indications are possible. One showing the track from the aircraft TO the station and the reciprocal showing the track to the aircraft FROM the station.

The aircraft does not have to be flying the heading of the radial inbound or outbound for the CDI to be centred. The receiver merely indicates that at that moment the aircraft is located on that radial regardless

of the aircraft heading. It could be flying along it or crossing it at an angle (Fig. 2-90). Thus, the VOR is sensitive to position but not to heading.

Homing to a Station

To *home* or fly directly to a VOR station, the following steps may be used:

1. Tune the receiver to the frequency printed on the chart and identify the station.
2. Rotate the OBS until the TO-FROM indicator shows a TO.
3. Continue rotating the OBS until the CDI is centred.
4. The reading under the OBS index is the magnetic track to the station. Turn the aeroplane to the heading indicated on the OBS. With the CDI centred the aeroplane will fly directly to the station. In a crosswind it may be necessary to repeat steps 3 and 4 to keep the CDI centred. This will result in the aircraft flying a curved path to the station.

Intercepting a VOR Radial TO a Station

Experienced pilots may use a number of methods to intercept a specific radial. However, to acquaint you with the use of the VOR, one basic technique will be presented (Fig. 2-91). Select the appropriate frequency and identify the VOR. Determine the reciprocal of the radial to be intercepted and set the corresponding number on the OBS. Check the TO-FROM indication. If the indication is FROM, you cannot readily intercept the desired radial from the present location. If the indication is TO, check the CDI. If the CDI indicates left, subtract 90 degrees from the OBS reading to determine the intercept heading. If the CDI indicates right, add 90 degrees to the OBS reading to determine the intercept heading. Fly the intercept heading until the CDI begins to move toward the centre, then reduce the intercept angle as necessary. Turn to the inbound heading as the CDI centres, and track to the station.

Intercepting a VOR Radial FROM a Station

If you want to intercept and fly along a radial away from the station, first select the appropriate frequency and identify the VOR. Next, determine the radial to be intercepted and set the corresponding number on the OBS. Check the TO-FROM indication. If the indication is TO, you cannot readily intercept the desired radial from your present location. If the indication is FROM, check the CDI. If the CDI indicates left, subtract 90

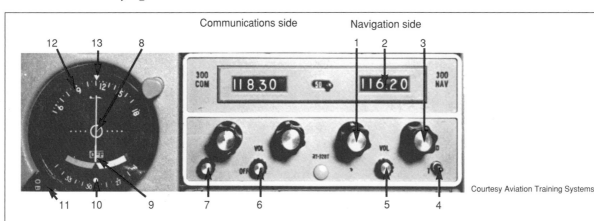

Communications side   Navigation side

Courtesy Aviation Training Systems

1 **Navigation Receiver Frequency Selector Knob** selects navigation receiver frequency in 1 MHz steps.
2 **Navigation Frequency Receiver Dial.**
3 **Navigation Receiver Fractional Frequency Selector Knob.**
4 **Ident Filter Switch** selects identifier. At "ID" position the filter is switched out of circuit and station identifier (Morse Code) signal is audible.
5 **Navigation Receiver Volume Control Knob** controls the volume of audio from the navigation receiver only.
6 **Off/On Volume Control Knob** turns the complete set on and controls the

volume of audio from the communication receiver.
7 **Squelch Control Knob** controls the communication receiver squelch circuit. Clockwise rotation increases background noise (decreases squelch action): counterclockwise rotation decreases background noise.
8 **Course Deviation Indicator (CDI)** indicates the course deviation from the selected omni bearing or localizer center-line.
9 **Off/To-From (Omni) Indicator** operates only with VOR or localizer signal. "Off" position (flag) indicates an unreliable

signal. When the "Off" position disappears, the indicator shows whether the selected course is "To" or "From" the station.
10 **Reciprocal Course Index** indicates the reciprocal of the selected VOR course.
11 **Course Selector (Omni Bearing Selector OBS Knob)** selects the desired course to or from a VOR station.
12 **Azimuth Dial.**
13 **Course Index** indicates the selected VOR course.

Figure 2-88   NAV/COM Radio with VOR

degrees from the OBS reading to determine the intercept heading. If the CDI indicates right, add 90 degrees to the OBS reading to determine the intercept heading.

Fly the intercept heading until the CDI begins to move toward the centre and begin to reduce the intercept angle as necessary. Turn to the outbound heading as the CDI centres and track from the station (Fig. 2-92).

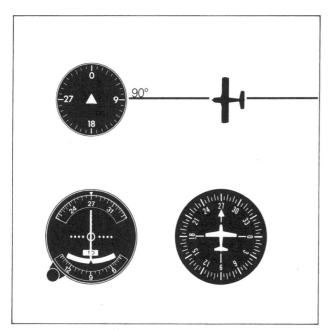

Figure 2-89   Aircraft on 090 Degree Radial Flying a Magnetic Track of 270 Degrees to the VOR

Correcting for Drift

Once you are established on a radial, it may be necessary to eliminate drift. If the wind direction is known, an allowance for drift may be made before the aircraft drifts off track. Fig. 2-93 shows the CDI indications that may be observed while following a radial. If the wind is not known, hold a heading identical to the radial being flown. After a short time any drift will be indicated by deflection of the CDI. The wind has drifted the aircraft in Fig. 2-93 to the right of the 090 degree radial. To reintercept, turn to a heading of 070 degrees and confirm that the CDI is moving toward the centre. An intercept angle of greater than 20 degrees may be required if the wind is strong. Hold the heading correction until the CDI centres. After the radial is intercepted, reduce the intercept angle to allow for drift. Subsequent left or right deflection of the CDI indicates insufficient or excessive drift correction and will require further heading changes to keep the CDI centred.

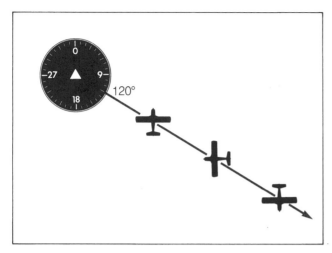

Figure 2-90   Aircraft on 120 Degree Radial Regardless of Heading

the half-way point or as depicted on the chart and set the inbound radial bearing on the OBS.

A Victor airway or air route joining two VOR stations does not always show the radials as exact reciprocal numbers because of chart convergence and magnetic variation. The CDI will not necessarily be centred when the new station is tuned, but it will indicate what direction to turn if necessary to intercept the new radial.

Determining a Fix

The position of an aircraft can be plotted by taking bearings from two or more VOR stations. The resulting lines of position are plotted on a navigational chart and the point where these lines intersect is called a *fix*. Because the aircraft is still moving ahead, the accuracy of the position will depend on how quickly the bearings are taken. If the aircraft has two VOR receivers this is not a problem. Fig. 2-94 shows how the aircraft's position may be plotted using lines of position.

Flying Victor Airways and Air Routes

Enroute Low Altitude (LO) charts have the VOR (Victor) airways or air routes clearly indicated, enabling you to readily determine the airway or air route to follow from one VOR station to another. When flying along a Victor Airway between two VOR stations, tune in the frequency of the station ahead at approximately

Station Passage

As the aircraft approaches a VOR station, the CDI becomes very sensitive and may show large deflections

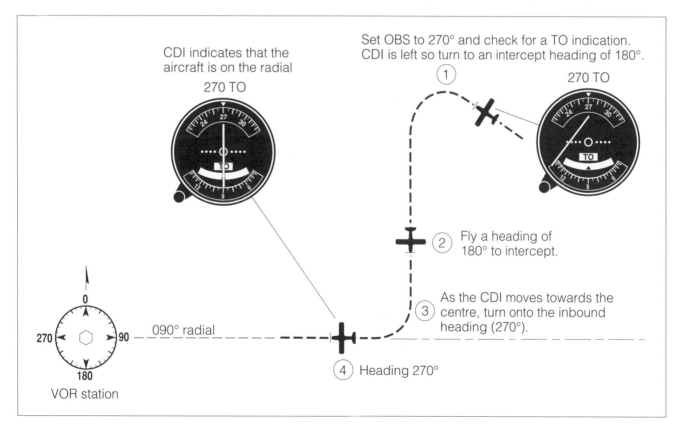

Figure 2-91   Intercepting a VOR Radial TO a Station

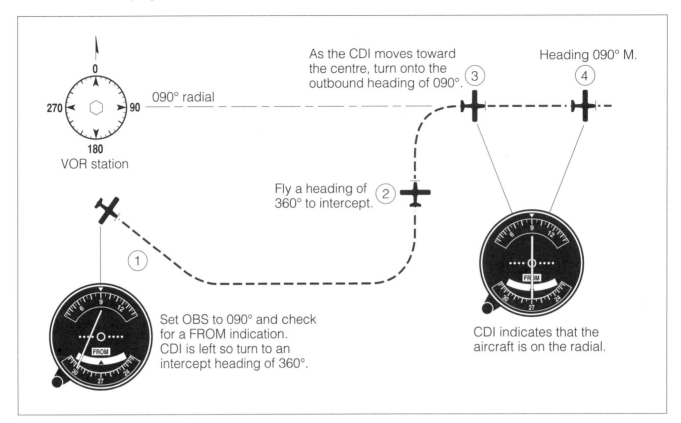

Figure 2-92   Intercepting a VOR Radial FROM a Station

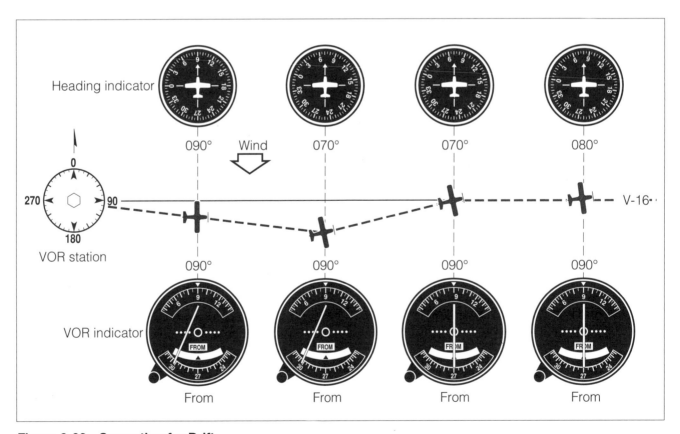

Figure 2-93   Correcting for Drift

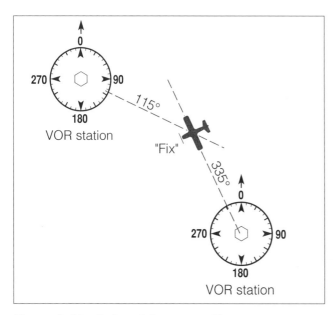

Figure 2-94 Determining a VOR Fix

although the aircraft is only a short distance away from the desired track. Hold the heading that kept you on track along the radial until the aircraft passes over the station. Station passage will be indicated by full deflection of the CDI to one side and then the other, with the TO-FROM indicator changing from TO to FROM. This is indicated in positions 1 and 2 of Fig. 2-95. If the aircraft is passing abeam the VOR station, the CDI and TO-FROM indicator will record the fact as shown in positions 3, 4, and 5 of Fig. 2-95.

If a change of direction is desired after station passage, select the outbound radial on the OBS and use the CDI as before to indicate the direction to fly to intercept the new radial. If you are close to the station, a 90 degree intercept angle will likely cause you to overshoot the radial. This can be avoided by selecting a shallower intercept angle.

Automatic Direction-Finder

The automatic direction-finder (ADF) is a low frequency radio receiver that can be used for reception of non-directional radio beacon (NDB) signals and commercial broadcast stations. The system has the ability to provide continuous relative bearings or magnetic bearings, or both, to any radio facility within the frequency range of 190 KHz to 1750 KHz.

When radio beacons are used as a navigational aid, the morse code identifier signals can be readily used to identify the beacon. However, if a radio broadcast station is used, it is essential to identify the station positively by listening to confirm station identification and to ensure that a back-up antenna site is not being used prior to relying on bearing indications.

ADF reception is not subject to line of sight transmissions. Reception range depends to a great degree on the strength of the broadcast station and atmospheric conditions. Compared with the static free qualities of VOR, radio static caused by lightning or any disturbance in the atmosphere is quite often a problem with the use of the ADF.

While there are other types of ADF displays, this text will discuss the display in which the longitudinal axis of the aircraft is parallel to a line passing through the zero index (0 degrees) and 180 degrees, as illustrated in Fig. 2-96. This is commonly known as the "fixed card" display. In this figure the ADF needle is pointing to a beacon that is 40 degrees to the right of the nose of the aircraft.

Terms and Definitions

Before using the automatic direction finder, you should understand the following definitions (Fig. 2-97).

Relative Bearing. The angle formed by the intersection of a line drawn through the centre line of the aircraft and a line drawn from the aircraft to the beacon. This angle is always measured clockwise from the nose of the aircraft. The relative bearing is indicated directly by the ADF needle when the beacon is tuned and the function selector knob is in the ADF position.

Magnetic Bearing. The angle formed by the intersection of a line drawn from the aircraft to the beacon and a line drawn from the aircraft to magnetic north. For an ADF radio with a fixed azimuth indicator, a magnetic bearing to the beacon is obtained by adding the relative bearing shown on the indicator to the magnetic heading of the aircraft. If the total is more than 360 degrees, 360 degrees is subtracted to obtain the magnetic bearing.

Reciprocal Bearing. The bearing plus or minus 180 degrees. Reciprocal bearings are used when plotting fixes. A reciprocal bearing (beacon-to-aircraft bearing) is obtained by adding or subtracting 180 degrees from the aircraft-to-beacon bearing. If the bearing is less than 180 degrees, 180 degrees is added to obtain the reciprocal bearing. If the bearing is more than 180 degrees, 180 degrees is subtracted to obtain the reciprocal bearing.

Position Line. When a bearing is taken on a beacon and plotted on a map, the resulting line is called a position line. Fig. 2-98 shows how a position line may be plotted.

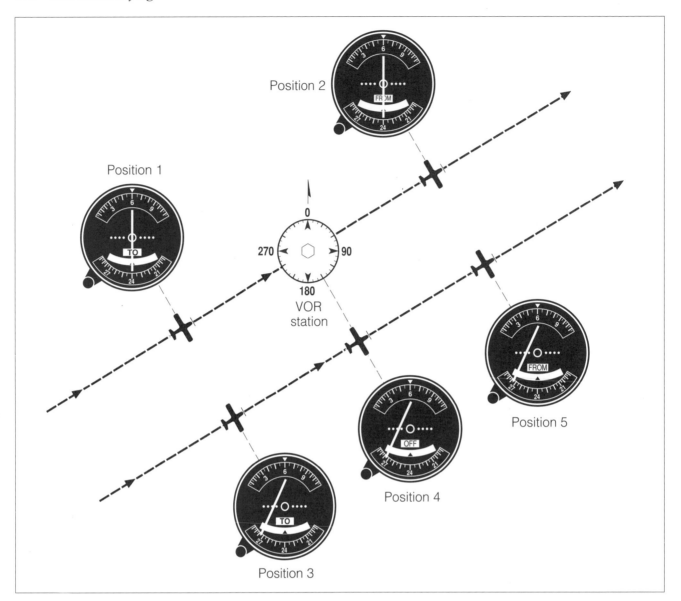

Figure 2-95   Station Passage

Homing to the Station

Homing to a beacon or station using the ADF in no wind conditions simply entails keeping the ADF needle on the 0 degree index. This method can be followed in a cross-wind condition but will result in a curved path as shown in Fig. 2-99.

Intercepting a Track to a Station

To intercept a desired ADF track to a station, first determine the aircraft's position relative to that track by turning to the same heading as the desired track. Next, turn 90 degrees from the parallel heading in the direction of the ADF needle. As you approach the desired track the needle will move toward either a 90

degree or a 270 degree relative bearing depending on whether you are left or right of the desired track. As the needle nears the wing-tip position, turn inbound on the desired track.

In Fig. 2-100 the aircraft is south of track with the ADF needle reading 030 degrees. For an intercept angle of 90 degrees, a heading of 015 degrees would be required. When established on the intercept heading, the ADF will indicate the station to be to the left and ahead of the wing-tip. When the angle opens to a relative bearing of 270 degrees, the aircraft is on the desired track and a turn should be made to the desired heading of 285 degrees.

If after turning to parallel the required track the ADF needle indicates a relative bearing between 090 degrees clockwise to 270 degrees, the beacon is behind the aeroplane. From this position you will be

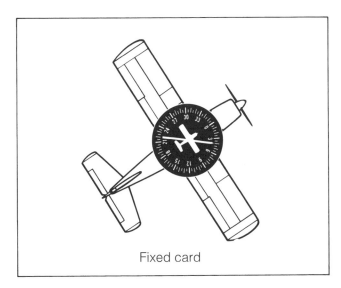

Figure 2-96   Fixed Card

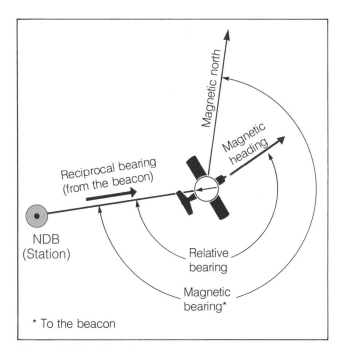

Figure 2-97   Bearings

unable to use the 90 degree intercept method to establish yourself on an inbound track of 285 degrees.

Intercepting a Track from a Station

To intercept an ADF track from a station, first determine the aircraft's position relative to that track by turning to the same heading as the desired track. Turn 90 degrees from the parallel heading in the direction of the ADF needle. As you approach the track the needle will move toward either a 90 degree or 270 degree relative bearing. As the needle moves to the wing-tip position, turn outbound on the desired track.

In Fig. 2-101 the aircraft is west of track, with the ADF needle reading 140 degrees. For an intercept angle of 90 degrees, a heading of 050 degrees would be required. When established on the intercept heading, the ADF will indicate the station to be to the right and ahead of the wing-tip. When the angle opens to a relative bearing of 090 degrees, the aircraft is on the desired track and a turn should be made to the desired heading of 320 degrees.

Correcting for Drift

Once you are established on a track, it may be necessary to eliminate drift. If the wind direction is

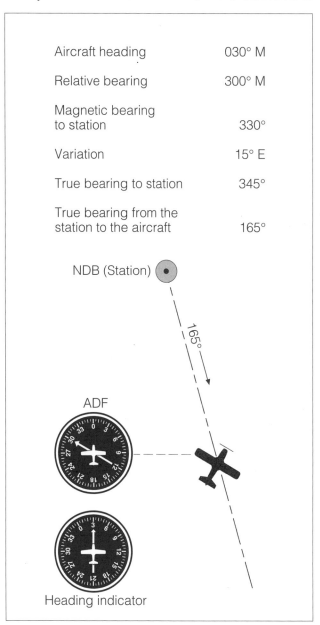

Aircraft heading	030° M
Relative bearing	300° M
Magnetic bearing to station	330°
Variation	15° E
True bearing to station	345°
True bearing from the station to the aircraft	165°

Figure 2-98   Position Line

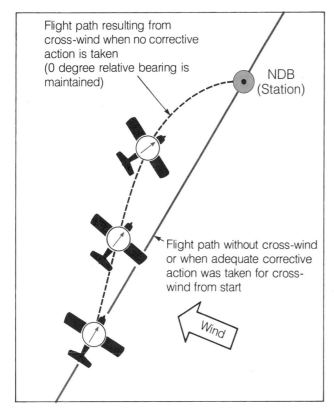

Flight path resulting from cross-wind when no corrective action is taken (0 degree relative bearing is maintained)

NDB (Station)

Flight path without cross-wind or when adequate corrective action was taken for cross-wind from start

Wind

Figure 2-99   Homing

Flying LF/MF Airways

A prescribed track joining two non-directional beacons is called a Low/Medium frequency LF/MF airway. LF/MF airways are marked on aeronautical charts and are based at 2,200 feet above ground level. An LF/MF airway joining two NDBs does not always show the tracks at each station as being exact reciprocal numbers due to the effect of chart convergence and magnetic variation.

When flying along an LF/MF airway between two NDBs, depending on the power output of the beacons, tune the frequency of the station ahead at approximately the half-way point. When the new station is tuned, the ADF indicator may not show you to be on track, but it will indicate the direction to turn if necessary to reintercept the airway.

Determining a Fix

As with VOR, when two or more magnetic bearings are taken on different NDBs or VOR/NDB combinations, and the resulting lines of position are plotted on a navigational chart, the point where they intersect will be the position of the aircraft at the time the bearings were taken. The intersection of these lines is called a fix.

Fig. 2-104 shows how the aircraft's position may be plotted.

Station Passage

Flight directly over the beacon can cause the needle to swing back and forth, possibly both clockwise and counter-clockwise, and finally to swing completely around to 180 degrees, indicating that the beacon is behind the aircraft.

When an aircraft approaches and passes abeam a beacon, the nearer the aircraft is to the beacon, the greater will be the number of degrees the needle is off 0 degrees. For example, if an aircraft is 1 mile to the right of a desired track 60 miles from the beacon, under no wind conditions, the ADF will show 1 degree off 0 degrees (359 degrees relative). At 30 miles it will be 2 degrees (358 degrees relative) and at 15 miles 4 degrees (356 degrees relative). This figure increases as the aircraft gets closer to the beacon, so that as the aircraft passes abeam the beacon at a distance of 1 mile, the needle would be 90 degrees off 0 degrees (270 degrees relative).

known, an allowance for drift may be made before the aircraft drifts off track. Any deviation of the ADF needle from the 0 or 180 degree index while maintaining the inbound or outbound heading shows that the aircraft is drifting. If it is drifting to the right as in Fig. 2-102, for example, turn to a heading of 070 degrees. Fly the new heading until the ADF needle moves clockwise to a position 20 degrees off the 0 degree index. After the track is reintercepted, divide the intercept angle by two and fly a heading of 080 degrees.

In Fig. 2-103 the aircraft is tracking away from the station and has drifted to the right of track. To reintercept, a turn to a heading of 070 degrees is made. This heading is maintained until the ADF needle moves to within 20 degrees of the 180 degree index. After the track is reintercepted, divide the intercept angle by two and fly a heading of 080 degrees.

The correct allowance for drift will be evident when the number of degrees the needle is placed off the 0 or 180 degree index remains constant, with a constant heading being shown on the heading indicator.

If you have not estimated the drift angle correctly, a change in bearing will again become apparent and further adjustments to the drift angle will have to be made until the ADF needle remains steady.

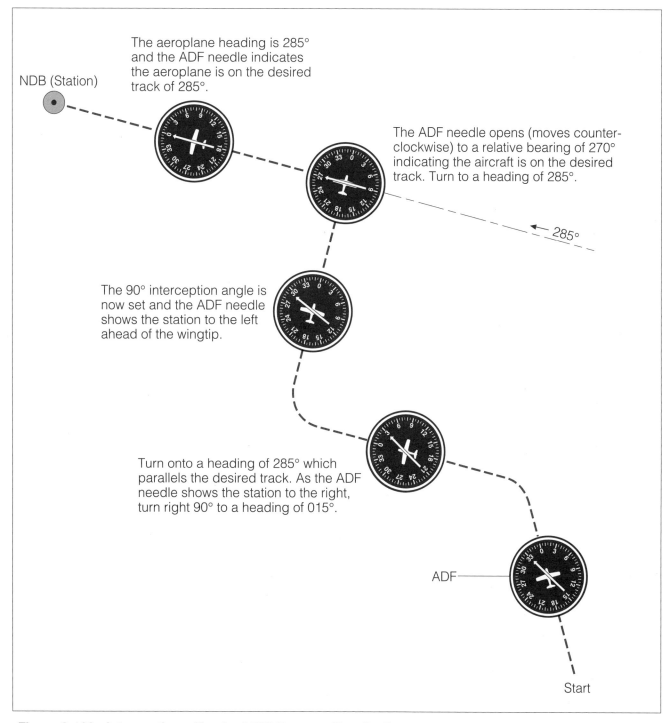

The aeroplane heading is 285° and the ADF needle indicates the aeroplane is on the desired track of 285°.

NDB (Station)

The ADF needle opens (moves counter-clockwise) to a relative bearing of 270° indicating the aircraft is on the desired track. Turn to a heading of 285°.

285°

The 90° interception angle is now set and the ADF needle shows the station to the left ahead of the wingtip.

Turn onto a heading of 285° which parallels the desired track. As the ADF needle shows the station to the right, turn right 90° to a heading of 015°.

ADF

Start

Figure 2-100 Intercepting a Track of 285 Degrees To a Station

Monitoring

The ADF volume should be left loud enough that it would become apparent if the beacon or aircraft receiver fails. Some aircraft receivers have a built-in fail safe mechanism, such as continuous rotation or rotation of the ADF needle to the wing-tip position, or an off flag, to indicate equipment failure. In this case, monitoring the station would not be as important.

Intercept Angles

Although a 90 degree intercept angle is the shortest route to a desired radial or track, it isn't the shortest route to the station. As you become more proficient

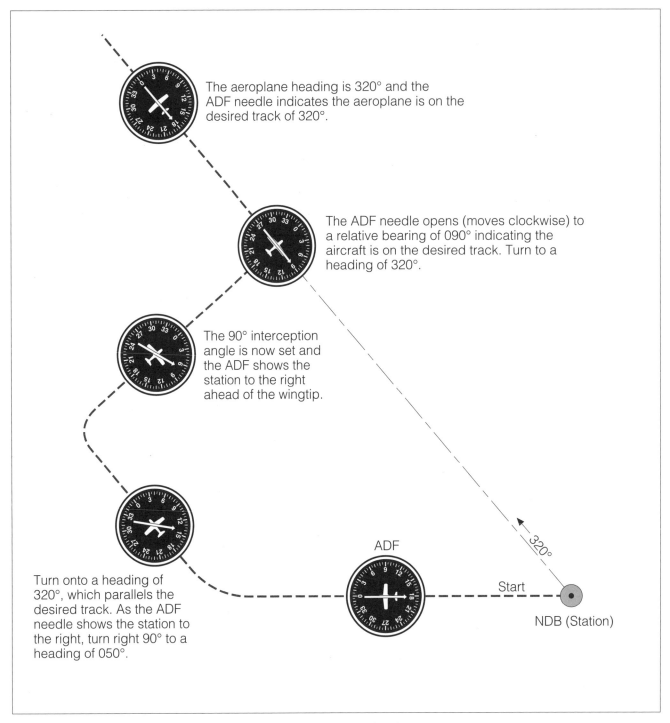

The aeroplane heading is 320° and the ADF needle indicates the aeroplane is on the desired track of 320°.

The ADF needle opens (moves clockwise) to a relative bearing of 090° indicating the aircraft is on the desired track. Turn to a heading of 320°.

The 90° interception angle is now set and the ADF shows the station to the right ahead of the wingtip.

Turn onto a heading of 320°, which parallels the desired track. As the ADF needle shows the station to the right, turn right 90° to a heading of 050°.

ADF

Start

NDB (Station)

320°

Figure 2-101   Intercepting a Track of 320 Degrees From a Station

in the basic intercept technique, your instructor will show you how to use shallower intercept angles that will take you more directly to your destination. In these cases, care must be taken to achieve the intercept *before* reaching the station. The use of a shallower intercept angle is also often used when intercepting near a station where the radials or tracks are close together, or when correcting for wind drift. When using an intercept angle less than 90 degrees, take care also that wind is not causing a rate of drift that exceeds your intercept angle.

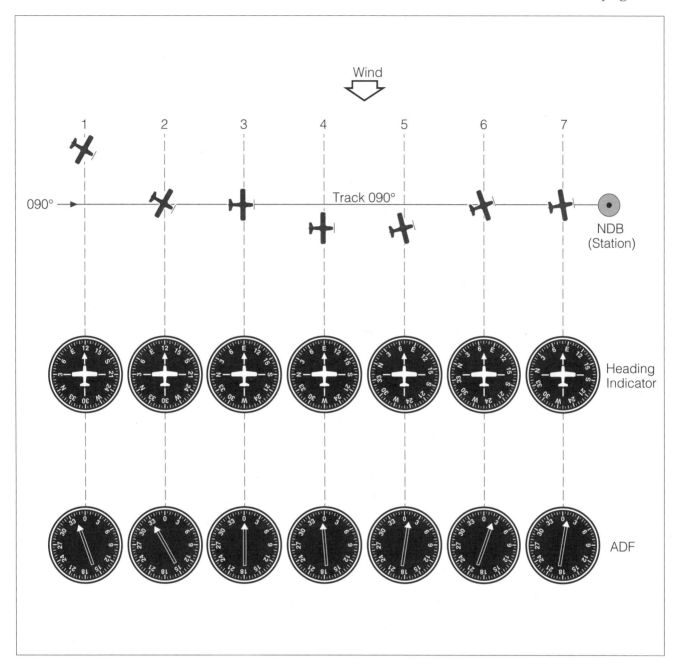

Figure 2-102   Correcting for Drift While Tracking Toward the Station

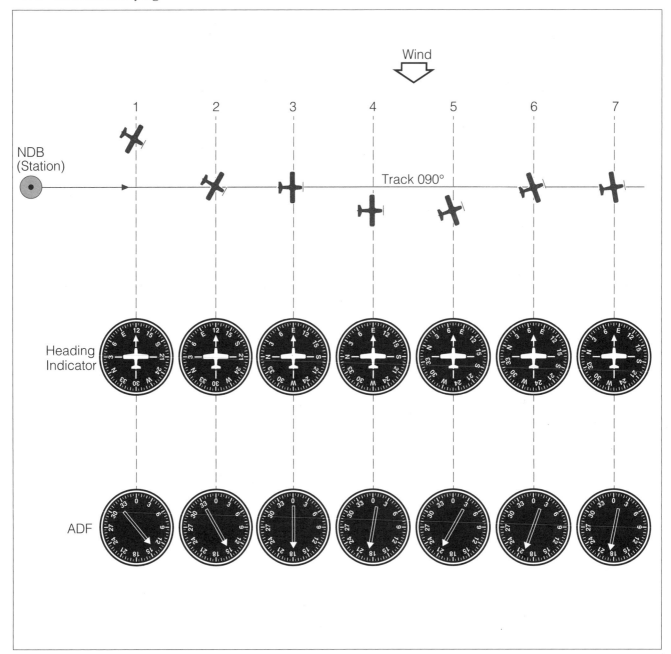

Figure 2-103　Correcting for Drift While Tracking Away From the Station

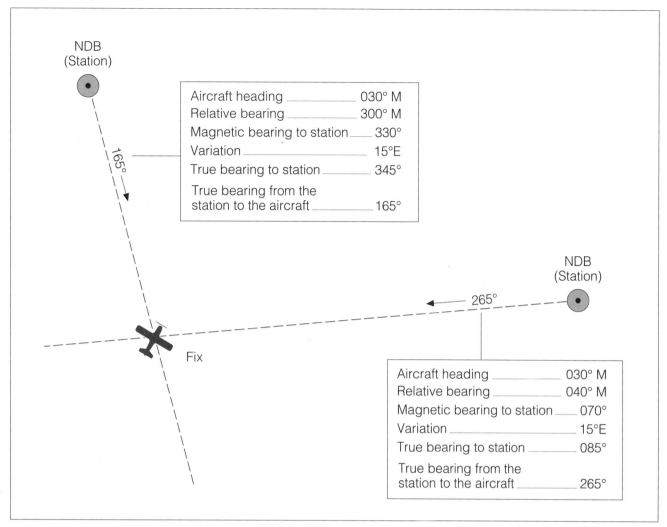

Figure 2-104 Determining a Fix

Night Flying

Night flying can be most enjoyable; in fact, many experienced pilots prefer it to flying by day. In summer, the lower temperatures at night make the air more dense, which improves aerodynamic and engine performance. Convection cloud tends to dissipate; therefore, air turbulence is much weaker and very often almost entirely absent. The air near the ground is generally more stable and good landings can be made with relatively less manipulation of the flight controls. Once accustomed to night flying, you will find that other aircraft in flight, which are generally less numerous than during the day, can be seen more readily.

Night flying does require that you readjust to a relatively different environment, especially outside the cockpit. Reference points such as the horizon, topographical features, and even the ground itself, all so vital in establishing aircraft attitudes by day, are indistinct, obscure, and sometimes invisible. Nevertheless, you will find that there is nothing mysterious or particularly difficult about night flying. The aircraft is flown by night in the same way as it is by day, though more frequent reference should be made to the instruments to verify attitude, airspeed, heading, etc. For this reason you must be adequately proficient in controlling the aircraft by reference to instruments.

You should become familiar with the airport lighting systems. Some of the names of these systems are: runway lights, runway threshold lights, runway approach lights, taxiway lights, taxiway entrance lights, airport rotating beacon, and obstruction lights. Various colours are used: threshold — green, runway edge — white, taxiway — blue, and obstruction — red. At many airports runway lights and approach lights can be varied in intensity. The intensity is usually controlled by the control tower or flight service and can often be varied at the pilot's request.

Before attempting a night flight, you must be thoroughly familiar with the operation of the aircraft's lighting system and its emergency equipment. Memorize the location of switches, circuit breakers, and fuses. Check that the required flashlight is working, that its batteries are sufficiently strong, and that it is within easy reach. Cockpit lighting must illuminate vital instruments and equipment satisfactorily, but should not create a glare that interferes with the pilot's outside vision. Position lights, sometimes referred to as navigation lights, must be checked for serviceability and operation. Do not forget the importance of the generator or alternator charging rate, since the load imposed both by the radio and the aircraft lighting system depend on it.

It may have been some time since you have had to use light signals directed at you by the control tower. Review these signals so that you know what to expect and do in the event of a radio failure.

The position lights of an aircraft are coloured and are located so that they are visible through certain angles for the express purpose of indicating the relative position of an aircraft and the general direction in which it is moving. It is important that you know how to interpret the position lights of another aircraft to determine whether there is any possibility of a collision.

Allow enough time after exposure to bright light for your eyes to become accustomed to darkness. Most people require about a half-hour in darkness for their eyes to achieve maximum adaptation.

During the day there is little possibility of flying into a cloud condition accidentally. On a dark, overcast night, however, it can be done easily. Be alert to the possibility of the existence of cloud in the area. At night it may be detected or suspected by the otherwise unwarranted disappearance of lights on the ground and by a red or green glow adjacent to the position lights of the aircraft.

Taxiing

In preparation for the flight carry out all normal daytime checks before taxiing out, and check your night flying equipment — such as landing lights, position lights, instrument and cabin lighting, the flashlight, and the alternator/generator charging rate.

Taxiing at night requires extra care compared with taxiing in the daytime, for the following reasons:

1. At night stationary lights are nearer than they appear to be, which makes judging distance difficult.
2. Speed is deceptive at night and there is a tendency to taxi an aircraft too fast. One reason for this is the lack of the customary visible ground objects which make speed apparent on the ground during the day.
3. Careful look-out is required to avoid obstructions. They are marked by red obstruction lights, which are sometimes mistaken for the lights of aircraft.
4. It is difficult to determine slight movement of the aircraft on the ground at night, and care should be exercised to prevent the aircraft from creeping forward during the run-up.

Take care not to shine your taxi/landing lights on other aircraft as a pilot's night vision may be impaired by the sudden bright light. Strobe lights should not be used on the ground for the same reason.

At the take-off position, if the aircraft is generator equipped, keep the engine running fast enough to keep the generator charging.

Take-off

Complete a normal pre-take-off check. Correct trim is important. It is also good practice to put the landing lights on, if the aircraft is equipped with them. Obtain take-off clearance (or take the required precautions at uncontrolled airports), then line up the aircraft on the runway in use.

The take-off is similar to the daytime take-off. Direction is maintained first by reference to the runway lighting and later by other lighted objects ahead. If landing lights are not used there may be some difficulty in judging the aircraft's attitude, so take care to ensure that the normal nose-up attitudes for take-off and climb are established. It is essential that a safe climbing speed with a positive rate of climb be achieved just after lift-off. Thus, any temptation to lift the aircraft off the ground prematurely must be resisted.

Because of difficulty in judging the pitch attitude, on some tail wheel aircraft the control column is allowed to remain neutral during the take-off run and the aircraft allowed to assume the flying attitude of its own accord. To guard against settling back to the ground after lift-off, the aircraft is climbed at a gentle nose-up attitude immediately it becomes airborne, until the desired climbing attitude may be safely assumed.

Supplement visual reference with instrument reference on take-off before losing sight of the runway lights. When established in the climb you may have to augment directional and pitch attitude control by reference to flight instruments, since visual references can often be obscured by the nose-up attitude. Do not start to turn until a height of 500 feet above ground has been gained, after which the aircraft should be put into a climbing turn in the direction of the circuit. Action in the event of engine failure after take-off is the same as by day, with the additional action of switching on the landing lights.

The Circuit

Regular circuit patterns are to be made, thus permitting completion of cockpit checks (and receipt of clearances) in accordance with the normal procedure on the various legs of the circuit. The circuit is flown principally by reference to other aircraft, aerodrome lighting, and lights on the ground.

Your instructor will point out other aircraft in flight and will show you how to space your circuit pattern to avoid crowding on approach.

Approach

A power assisted approach is normally used, but a low flat approach should be avoided. It is an important principle of night flying that pilots be able to complete safe approaches and landings by reference to the runway lighting only. The normal aid to judgement is the appearance of the runway lighting as seen after turning in on the final approach. If the approach path is correct, the lights will appear to remain equidistant (longitudinally). If you are overshooting, the distance between lights appears to increase; if undershooting, the distance appears to decrease. The aim, therefore, is to regulate the approach path so as to maintain the runway lights in the correct perspective (Fig. 2-105). Ideally, the approach should be gauged so that the flare occurs over the beginning of the runway lighting.

Landing

The landing at night is made by visual reference to the runway lights. The appearance of the ground is deceptive: never attempt to refer to it as you do by day.

The normal but not invariable effect of night conditions on the pilot is to induce a tendency to flare too high. If you experience this difficulty early in

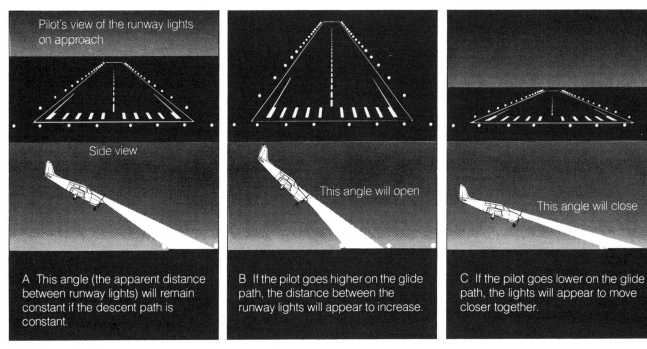

Pilot's view of the runway lights on approach

Side view

A This angle (the apparent distance between runway lights) will remain constant if the descent path is constant.

This angle will open

B If the pilot goes higher on the glide path, the distance between the runway lights will appear to increase.

This angle will close

C If the pilot goes lower on the glide path, the lights will appear to move closer together.

Figure 2-105   Runway Approach at Night

instruction, it may be advisable to keep some power on until the touchdown is completed.

When the aircraft's landing lights are used you should not look directly down the beam, but ahead of it and slightly to one side, in order to avoid losing perspective. Remember that the flare is still gauged by reference to the runway lights.

Executing a missed approach by night requires no special technique but is conducted in the same manner as by day. Reference to the flight instruments should be made before losing sight of the runway lights to augment directional and pitch control as discussed in the take-off procedures.

At night as during the day, you must keep alert for other air traffic and avoid keeping your head in the cockpit to read instruments for too long a period. As you gain experience in flying night circuits, you should become familiar with the correct position of the flight instrument needles for various flight attitudes and airspeeds so that it is not necessary to try to read the actual figures, and a glance is sufficient to determine that reasonable accuracy is being achieved.

Cross-Country Flying by Night

The principles of pilot navigation by night are basically similar to those applicable by day, except that map reading at night calls for special techniques. The aircraft is navigated according to a predetermined flight plan, corrected from time to time by radio navigation and reliable visual aids.

The following points call for special consideration when flying cross country at night:

1. A weather briefing and detailed flight planning are essential.
2. Identifiable lighted landmarks or large lakes and rivers are easier to see at night.
3. Compass headings should be accurately maintained and corrections made only when the position, fixed by check-points or by radio aids, is absolutely certain.
4. Accuracy in time keeping is essential.

The route for initial night pilot navigation practice should be carefully chosen to include several landmarks that can be identified unmistakably at night. The feasibility of map reading will depend mainly on the weather and the moon. Ground features show up better when viewed against the moon. Aerodrome beacons are very useful fixes, but guard against the possibility of large errors when judging distance to or from them. Avoid depending on small lights on the ground for fixes; the scattered lights around a small community can give the impression of a much larger town. At all times be aware of the approximate bearing and distance of a known prominent feature that you can divert to easily should anything occur to make continuation of the flight impossible or impracticable.

Airport Lighting

The floodlights illuminating most terminal building aprons are in many cases semi-blinding and cannot be dimmed or turned off except in emergencies. When taxiing in floodlit areas use extreme caution, since persons, vehicles, and other objects tend to "shadow" out or become invisible in these areas.

Blue lights are used to delineate taxiways and are necessarily of very low intensity. To assist aircraft exiting from a lighted runway, the exit is generally identified by two blue lights on each side of the taxiway exit.

Runway and approach lighting systems are not restricted to night-time use. If visibility is poor, a landing is being made into the sun, or any other factor affects the safety of the aircraft in this regard, do not hesitate to ask the control tower or the agency responsible to turn on the runway and approach lights. These lights may also be adjustable in intensity; ask for lighting that is best suited to the circumstances.

Some aerodromes may use retro-reflective markers in place of lights to mark the edge of runways. These markers are approved for use on runways at registered aerodromes only; however, they may be used as a substitute for edge lighting on taxiway or apron areas at some certified airports.

Retro-reflective markers are positioned such that *when the aircraft is lined up on final* they will provide the pilot with the same visual presentation as normal runway lighting. A fixed white light or strobe light is installed at each end of the runway to assist pilots in locating the aerodrome and aligning the aircraft with the runway.

Retro-reflective markers must be capable of reflecting the aircraft landing lights so that they are visible for a distance of 2 NM. Be cautioned that the reflective capabilities of retro-reflective markers are greatly affected by the condition of the aircraft landing lights and the prevailing visibility. Therefore, as part of pre-flight planning to an aerodrome using retro-reflective markers, be sure to exercise added caution in checking the serviceability of your aircraft landing lights and making provision for an alternate airport with lighting in case of an aircraft landing light failure.

Floatplanes

Most of the civil primary flight training aircraft that use water instead of land as a take-off and landing surface are conventional landplanes equipped with two floats instead of wheels. All aircraft capable of taking off or landing on water are termed seaplanes, including flying boats. For the purpose of this text, the subject aircraft is a floatplane, since we do not intend to include any of the operating procedures applicable solely to flying boats. For the most part, the subject will be referred to as an aircraft.

In the air a floatplane acts much like a landplane. It does not require as much use of the ailerons in a sideslip and tends to be less stable directionally than a landplane. Otherwise, any normal manoeuvre that can be performed by a landplane can also be performed by a floatplane. Accordingly, no special instructions will be given here concerning operating of the aircraft in the air. The same applies, to a large extent, to familiarization with the aircraft itself.

The techniques for handling and manoeuvring a floatplane on the water are very different from those for handling a landplane on land. Besides airmanship the pilot must also acquire and apply knowledge in seamanship. A floatplane has no brakes and is affected by both wind and water currents. Whether the engine is stopped or running, left to its own devices the aircraft will always turn into the wind. Therefore, the stronger the wind the more difficult it is to manoeuvre a floatplane. Despite the additional problems, floatplanes can offer much pleasure and operational versatility. In most areas of North America the floatplane has countless "aerodromes" with "runways" of unlimited length.

Performance

Due to the weight of the floats, the useful load of a floatplane is normally less than that of the same aircraft on wheels. The rate of climb and cruising speed

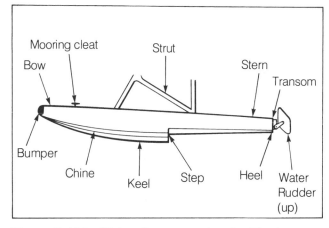

Figure 2-106   Major Components of a Float

are also reduced due to increased drag produced by the floats and the float attachments.

Seamanship

Terms

Under Way.   An aircraft that is not moored or fastened to any fixed object on the land or in the water is under way. A floatplane under way may be moving forward, backward, or sideways upon the surface of the water with its engine running or stopped, and may be taxiing, sailing, or stationary. Under way presumes that a properly authorized person is at the controls of the aircraft.

Sailing.   A floatplane that is under way and is being manoeuvred backward or sideways solely with the wind or water currents providing the necessary force is sailing. The engine may be running or stopped.

Bilgewater. All floats leak water to a varying degree, adding to the overall weight of the aircraft and diminishing buoyancy. The water is removed by means of a bilge pump. It is essential that this be done as often as necessary to keep the floats dry, especially before the first take-off of the day or when the floats have been subject to a hard landing or other abuse. The centre of gravity will also be affected by water in the floats.

Stationary. A floatplane under way that is being held in one position against wind or water currents by means of engine thrust is not taxiing, but is said to be *stationary*.

Taxiing. A floatplane under way that is being manoeuvred in a forward direction by means of engine thrust is *taxiing*.

Lines. Lengths of hemp, manila, or nylon rope used for mooring a floatplane or securing it to a dock are called *lines*.

Bridle. A Y-shaped configuration of lines used when mooring a floatplane to a buoy or when using an anchor. To allow a floatplane to weathercock properly when moored to a buoy or anchored, the bridle must be equilaterally secured to the bow cleats of each float. The longer the single line of the bridle the less the aircraft will tend to drag the anchor or mooring (Fig. 2-107).

Equipment on Board

In addition to the items carried by any aircraft to meet operational requirements, a floatplane should also have on board the following equipment, in serviceable condition:

1. An approved and readily accessible life-jacket for each occupant.
2. Two or more 25 foot mooring lines.
3. A float bilge pump.
4. One or more paddles.
5. An anchor with its own 50 foot line.
6. A device for filtering gasoline when away from base.

General Considerations

Footwear. Floatplane pilots should wear shoes or boots that afford a good grip on the normally wet surface of the floats. Footwear with nails or cleats on the soles may scratch or damage the surface of the floats.

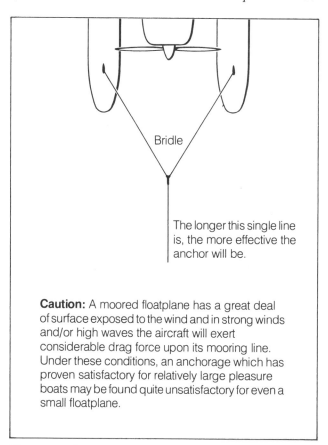

The longer this single line is, the more effective the anchor will be.

Caution: A moored floatplane has a great deal of surface exposed to the wind and in strong winds and/or high waves the aircraft will exert considerable drag force upon its mooring line. Under these conditions, an anchorage which has proven satisfactory for relatively large pleasure boats may be found quite unsatisfactory for even a small floatplane.

Figure 2-107   Bridle

Starting the Engine

Many variables apply to the actions immediately preceding the actual starting of the engine. If no helper is available to hold the aircraft until the engine is started and you are ready to taxi, then the aircraft should be suitably restrained and not released until you are seated. Release in this case can be made by releasing a rope that is threaded through a fitting or around a strut. If it is necessary to allow the aircraft to drift away from its mooring, then surrounding obstructions such as other aircraft, trees, rocks, and piers will dictate the procedure to be followed. If you face this situation with a clear plan, such as releasing an anchor or paddling, should the engine fail to start there is less likelihood of damaging the aircraft. If the engine has to be started by hand the propeller must be swung from behind. This is done by standing on the right float with a firm grip taken on a convenient strut. Usually only the smaller engines can be started in this manner. The *Canadian Aviation Regulation* governing engine starting and engines left running applies to floatplanes as well as landplanes. A floatplane has no brakes and will begin to move immediately when the engine starts, unless suitably restrained.

Water Currents

Special operating considerations may be necessary when water currents, such as may occur in a river or in tidal action, exceed 5 KT, especially in areas where obstructions or other hazards are present.

1. When turning from downwind to upwind for take-off:
   (a) if taxiing against the current, begin the turn beyond the intended take-off point; or
   (b) if taxiing with the current, begin the turn before the intended take-off point.
2. The best conditions for take-off occur when the take-off is made with the current (in the direction the current is moving) and into the wind.
3. The best conditions for landing occur when the landing is made against the current and into wind.

The Flight

Pre-Flight Inspection

In addition to the usual items inspected in a landplane during a pre-flight inspection, some important items in a floatplane warrant careful attention. The float compartments must be inspected for water, pumped out, and closed. The load penalty from water leaking into the floats can mean increased take-off distance, which could produce an accident. The condition of the float struts, brace wires, attachments, fittings, cables, water rudders, and paddle attachments should be ascertained.

Passengers

As a general rule, passengers should not be permitted out of the cabin while the propeller is turning. If passenger assistance is required, a thorough briefing on procedures and hazards must be given. Many a passenger has been hit by the idling propeller or fallen into the water while trying to assist a pilot to secure or release a line. Therefore, ensure that passengers stay seated until the aircraft is secured when docking or departing.

Taxiing

The proposed taxi path should be planned in advance if obstructions exist, as wind strength, river current, tidal action, or a combination of these factors will dictate your actions while proceeding to the take-off area. When preparing to land at an unknown area, select the taxi path from the air when underwater hazards can be seen and other obstructions can be noted.

To aid in turning while taxiing at slow speeds, floatplanes have water rudders, hinged to the transom of one or both floats. They are linked to the normal rudder control system and may be retracted or lowered by the pilot from the cockpit. Water rudders are most effective at low speeds in comparatively calm water; at high speeds the pressure of the water tends to kick them up into the retracted position.

A floatplane has three taxiing modes, known as *idling*, *sailing*, and *on the step*. In the first mode the elevator control should be held all the way back so that the propeller does not strike spray developed by the bows of the floats or wave action, since water striking a rotating propeller can inflict severe damage to the propeller and its components. Except when special wind and water conditions prevail, or there are other extenuating circumstances, experienced floatplane pilots consider it normal practice to taxi by means of a combination of the idling mode and sailing.

Idling

In the idling mode (Fig. 2-108), the speed of the aircraft through the water is approximately 8 KT or less, and the aircraft's attitude is about the same as when it is at rest on the water. Spend as much time as possible initially taxiing in the idling mode to familiarize yourself with the action of the water rudders. This practice is best conducted when the wind speed is below 10 KT and the water relatively calm.

Great care must be taken when making turns on the water, especially at high speed or in a strong wind. Floatplanes constantly endeavour to turn into the wind (weathercock) when being taxied across wind or downwind. Consequently, when countering control pressures are relaxed, the aircraft will swing abruptly into wind. Centrifugal force tends to make the aircraft roll toward the outside of the turn and the wind

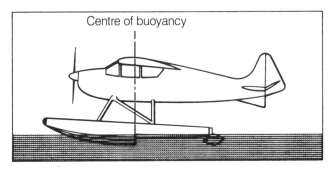

Figure 2-108   Idle Taxi

striking the side of the aircraft further aggravates the rolling tendency. If you turn abruptly when taxiing downwind, the combined action of the two forces can be sufficient to overturn the aircraft. Additionally, the more the aircraft heels over the greater the lifting effect of the wing exposed to the wind on the windward side and the less on the wing on the leeward side.

To make a turn into wind, simply neutralize the rudder, unless you are taxiing directly downwind; in that case a slight amount of rudder should be applied in the direction in which a turn is desired. As soon as the turn begins, neutralize the rudder. If the wind is strong, apply opposite rudder to slow the turn down.

Sailing

Sailing is a procedure used to position or manoeuvre a floatplane in, or to, an area where normal manoeuvring is inadvisable because of congestion or unfavourable wind or water conditions. With even the lightest breeze, a floatplane may be sailed into a very cramped space. If there is absolutely no wind the aircraft can easily be manoeuvred with a paddle. Water rudders must be retracted when sailing.

In a light wind with the engine stopped, a floatplane moves backward in the direction in which the tail is pointed. In a stronger wind with the engine idling, movement is usually backward and toward the side to which the nose is pointed. Move the tail or nose in the desired direction by coarse application of rudder. Additional assistance can be acquired by fullscale deflection of the ailerons. When sailing with engine on, use the air rudder and aileron drag to steer the aircraft. A floatplane can travel as much as 45 degrees to the wind in this manner.

To sail directly backward, merely centralize all controls. Additional "sail" effect may be acquired by lowering flaps, lowering elevators, and opening cabin doors.

Care should be exercised and relative effects assessed where strong water currents or tidal action are present, since it is quite possible that these forces may offset the effect of the wind to varying degrees. When sailing near obstructions with the engine stopped, it is important to be able to restart the engine instantly, so that backward motion of the aircraft may be arrested immediately to avoid collision.

In the case of an engine that may not restart readily and must be left running should it be needed to arrest backward motion, its forward thrust while sailing may be reduced by allowing it to run on one magneto only. Carburettor heat may also be used to reduce idling thrust. Do not operate the engine this way for too long at any one time. Once moored or docked, allow it to run on both magnetos for a short while before shutting it down.

Taxiing on the Step

Due to the higher speeds and other considerations involved, taxiing on the step requires considerable skill and experience, together with a good knowledge of water obstructions or other hazards in the locality. To gain solo experience, carry out your initial practice on smooth water with light winds and in areas you know well.

The aircraft is placed on the step by holding the elevator control fully back and applying full power. As the power is applied the nose will begin to pitch up and the aircraft will begin accelerating. You will notice that at some point the nose will rise no farther and there will be no further acceleration. When this point is reached, ease the control column forward and place the nose of the aircraft in an attitude slightly above the attitude the aircraft would be in at rest on the water (Fig. 2-109). As this is done the aircraft will begin to accelerate noticeably again.

As the acceleration will be fairly rapid, the power must be reduced in order to prevent the aircraft from becoming airborne. About 65 percent power should be sufficient for the procedure.

Should the nose of the aircraft begin to pitch up and down (a motion referred to as *porpoising*), it must be stopped immediately, as the oscillation will increase rapidly and the aircraft may become uncontrollable. The safest course of action is to close the throttle and hold the control column fully back, allowing the aircraft to return to idle taxi. Providing you recognize the porpoising action in the early stages, you can stop it by applying a back pressure on the control column as the nose pitches up.

Turns may be made on the step but they should be very gentle, and only a few degrees at a time, until you are thoroughly familiar with a particular floatplane. The aircraft is moving in excess of 25 KT over the water and the centrifugal force in too sharp a turn can easily capsize it. With certain wind and water conditions it is unsafe to execute a step turn under any circumstances. For example, if the wind speed is considerable, say over 20 KT, and the waves high, as the aircraft turns broadside to the wind the upwind float may be lifted by the crest of a wave while the downwind float is in a trough between waves. Under these conditions, if a turn is well established the aircraft is in danger of capsizing. Hence, if tendencies to heel over are evident at or near the start of the turn, throttle right back and apply rudder to stop weathercocking. Turns beyond 45 degrees on the step require a high degree of skill and experienced assessment of all existing circumstances and conditions. If any doubt exists as to the safety of this type of turn for a given condition, manoeuvre the aircraft by some other method, such as sailing. (The water rudders must be in the "up" position while taxiing on the step.)

In addition to the three methods of taxiing previously discussed, it is possible to taxi the aircraft in what is called the nose-up mode (Fig. 2-110). It must be clearly understood, however, that except for engine run-up, any nose-up taxiing should only be attempted by highly qualified seaplane pilots, due to the danger of upset. A general rule of thumb is that if the nose-up mode is necessary to turn downwind while taxiing in a high wind, then the average seaplane pilot should not be out there in the first place. If you are faced with such a situation, sailing backward is recommended.

To enter this mode hold the elevator control fully back and apply about half maximum RPM for the aircraft. Be careful to hold the control column fully back during this procedure to keep the propeller from being damaged from the spray should the nose get too low. Taxiing in this mode may be necessary in rough water and when turning downwind in high wind conditions.

There is considerably more float "side area" ahead of the centre of buoyancy than aft of it. Therefore, when taxiing cross-wind in the nose-up mode many aircraft tend to turn downwind instead of following the normal tendency to turn into the wind. This is why it is often necessary to adopt the nose-up mode when attempting to turn out of a high wind. It is possible to use power to augment rudder since opening the throttle increases the speed of the aircraft, causing the nose to rise higher, which exposes more float area and thereby increases the tendency to turn downwind. Conversely, reducing power decreases speed, lowers the nose, and allows the aircraft to turn into wind.

Taxiing with the nose up should be limited to short periods of time, as the engine can very quickly become overheated. In addition, because of the relatively high speed and limited forward view, be very careful to ensure that the path ahead is clear.

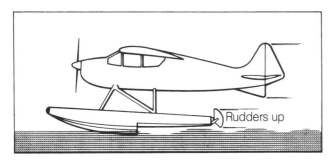

Figure 2-109 Taxiing on the Step

Mooring and Docking

When approaching a dock or mooring point, observe it closely for obstructions and study the possible effects of wind and water currents. Left to its own devices, a floatplane will always point into wind, and it can always be turned into wind without difficulty.

Always have a suitable line ready when approaching a dock or other mooring if no shore assistance is available. Taxi speed can be reduced by operating the engine on one magneto and/or using carburettor heat. Open the doors and release the seat-belts. Brief the passengers on what is planned, and if necessary place them for easy pilot exit. As the aircraft nears the dock and it appears that inertia will carry the aircraft the remaining distance, shut the engine down, leave the aircraft, secure a line to a strut, and when sufficiently close, step off with rope in hand. At the time the engine is shut down, be prepared for weathercocking due to loss of slipstream. Protect the aircraft from damage by using one foot to cushion the contact with the dock, and secure the line. In high wind conditions, if the line cannot be secured quickly enough, be prepared to jump back on to the float, restart the engine, and try again.

A seaplane ramp is a wide sloping surface, often of wood, with its lower extremity under the water. It is used for bringing seaplanes out of the water for sundry reasons, including routine docking. Other docking forms include piers, rafts, and buoys. The technique for approaching any of these varies, but if possible all of them should be approached into wind at a slow speed, since under this condition you have maximum control.

In the case of a raft (float) moored some distance from shore, even if the wind is blowing shoreward it is sometimes possible to taxi past the raft, turn, and then approach into wind. The same is true of a pier, since three sides are available for approach.

When "docking" an aircraft on a natural beach, ascertain the nature of the shore before contact is made. If it is rocky there is danger of damaging the floats, especially if waves rock the aircraft up and down. Sandy beaches are the best, but even these will wear off paint and protective coatings if there is wave action. With an onshore wind the best approach to a beach is to sail the aircraft backward to it with the water rudders up. This has the added advantage of not having to wade into the water to turn the aircraft around for departure.

If the wind is offshore, approach slowly, checking for submerged obstructions or obstacles that could damage the wings or tail. If the wind is onshore and very light, the same type of approach can be made. However, if the wind strength dictates, raise the water rudders and sail backward, watching for possible damage from obstacles on the shore. If the wind is parallel to the shore, taxi close until opposite the beaching point, then use the engine to turn the nose into shore and beach as soon as possible. Alternatively, taxi downwind and close the throttle when in a position where the weathercock action will face the aircraft for a close approach. In high winds, always use a helper on shore.

Anchoring to a Buoy

Approach the buoy from into wind at minimal speed. When inertia will carry the aircraft to the buoy, shut down the engine, and exit the aircraft with rope in hand. Secure the aircraft with a bridle or two ropes, one to each bow cleat.

Using an Anchor

Select a location, taking into account other aircraft or boat traffic, river currents, tide, wind speed, wave size, and depth of water. As a general rule, the anchor line should allow for 10 feet of length for every foot of water. Always ensure that the anchor is holding before leaving the aircraft, and if wind speed increases, return to the aircraft to ensure its safety.

Leaving a Dock or Mooring

Departing from a pier or raft (float) presents no real problem. It is desirable to have an assistant hold the aircraft pointed toward open water until you start the engine. If no assistance is available and the bows of the floats are headed against the pier or raft, cast off and allow the aircraft to drift back far enough to make a turn without striking the pier or raft before starting the engine. When an aircraft is cast off and allowed to drift, the engine may not start readily, so always keep in mind the possibility of drifting into obstructions or obstacles.

Departing from a Buoy

In calm conditions it is possible for a float to be directly over a buoy anchor. Damage could result if the aircraft is in shallow water and you step on to the float. The buoy anchor could also do damage when moving away, so always exercise care in shallow water. Always position the buoy for departure so that it is at the side rather than between the floats.

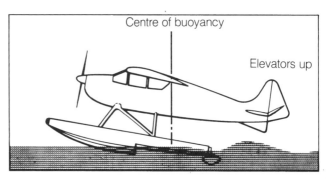

Figure 2-110   Nose-up Taxiing

Departing from a Beach

Depending on the strength of the wind and its direction, push the aircraft out from shore with a paddle or sail backward. Alternatively, face the aircraft toward the open water, start the engine, and put the water rudders down as soon as possible.

Taking Off

Take-off training should begin when there is enough breeze to make small waves but not enough to produce white caps. At the take-off position the water rudders should be placed in the "up" position, and the intended take-off path carefully studied to make sure that it is clear and will remain clear. Operators of pleasure boats, who are unaware of the operating requirements of aircraft, are likely to move directly into the path of a floatplane while it is taking off.

Unlike the landplane, the floatplane ordinarily takes off and lands in public areas. It therefore always faces the possibility of encountering partially submerged floating objects and swimmers that are difficult to see, as well as various types of watercraft. When taxiing into take-off position it is often advisable to taxi along the intended take-off path and check for floating obstacles and obstructions.

The take-off on floats is similar to a soft field take-off in a landplane. When the aircraft is settled into wind, hold the control column hard back, and smoothly apply take-off power, then allow the aircraft to progress through the nose-up mode into the planing mode according to the procedures already discussed. When the aircraft is in the planing mode (on the step) at full power, back pressure should be exerted on the control column. Too much back pressure will cause the heels of the floats to dig into the water and create a drag that will impede the take-off. Conversely, if insufficient pressure is exerted, the forward part of the bottom of the float will remain in the water and create undesirable drag. A common error is attempting to "drag" the aircraft out of the water, without realizing that the heels of the floats will be forced down into the water at a much lower angle than is required to put the tail wheel of a landplane on the ground. Experience will determine the best take-off attitude for each aircraft. If held at this attitude, the aircraft will take to the air smoothly and with adequate flying speed.

The importance of the proper use of flaps cannot be ignored. As a general rule in selecting the take-off path, when you feel you have sufficient distance, double it. The climb-out path must be planned prior to starting the take-off, with hills, valleys, and downdrafts being taken into account.

Rough Water Take-offs

When taking off on rough water, apply take-off power as the bows of the floats are rising on a wave. This prevents the nose of the floats from digging into the water and helps keep spray away from the propeller. Throughout a rough water take-off, hold the controls somewhat farther back than in smooth water, to hold the bows of the floats well above the surface. Once in the planing mode, the aircraft will begin to bounce from wave to wave, and each time it strikes a wave the bows will tend to rise. If nothing is done to correct this, the impact with each successive wave will be more severe. As the nose starts rising, exert forward pressure on the controls, then apply back pressure just before the aircraft strikes the next wave. It is important to exert back pressure at the correct instant, otherwise the bows of the floats may be pushed underwater, capsizing the aircraft, nose first. Accurate timing and quick reactions are essential. Fortunately, it usually follows that if there is wind enough to cause waves the aircraft will become airborne quickly.

The worst rough water condition occurs when a strong current runs against the wind. For example, if the current is running at 10 KT and the wind speed is 15 KT, the relative speed between wind and the water is 25 KT. The waves will be as high as those produced in still water by a wind of 25 KT.

Effect of Winds

Besides the obvious effects on taxiing, high winds can impose limitations on float operations, depending on the size of the aircraft and the experience of the pilot. Take-off and landing paths in sheltered areas are desirable, as open water can be rendered unusable by high winds. Learn to "read" the wind strength from the air by observing wave action, wind lanes (streaks upon the water), and other indications, before selecting a landing area.

Glassy Water Take-off

The take-off condition that may require more skill than any other occurs with the combination of a hot sultry day, calm wind, glassy water, and a fully loaded aircraft. Such a condition calls for timing, co-ordination of controls, and practice. The take-off run required will be longer, and it will take longer to get the aircraft on the step than under normal conditions. Once the throttle is opened with the control column held back, and you are quite sure that the nose of the aircraft has come up to as high an attitude as possible, ease the control column forward to roll the aircraft onto the step. Make sure that it is on the step and is accelerating and planing in the correct attitude. Glassy water is "sticky." When aileron control is effective, one float is gently lifted out of the water (thereby reducing water resistance); at the same time the nose is raised slightly, held there for one or two seconds, and the aircraft brought to level flight position. You will then find the aircraft is airborne. Delicate handling is necessary for the first few seconds as the nose may assume too high an attitude, due to the sudden unsticking from the water. If the nose of the aircraft is levelled too soon it will contact the water again. It is impossible to judge vertical distance from glassy water. Leave the flaps until plenty of height is gained, then raise them slowly.

Rocking onto the Step

There may be occasions, with a heavily loaded aircraft, hot weather or high altitude, and glassy water conditions, when it is necessary to rock the aircraft onto the step. *This should only be attempted if it is absolutely necessary.* Rocking the aircraft is in fact a controlled porpoise started by pulling the control column right back and then releasing it. The nose of the aircraft rises and falls immediately the control column is pulled back again. This procedure is repeated until the nose of the aircraft reaches a maximum height. If the control column is then pushed well ahead and held there, the aircraft is jumped onto the step. The best planing attitude is then carefully controlled with the control column. Allow the speed to build up as much as possible and follow glassy water take-off procedure.

It is difficult to give a written explanation of the proper timing for rocking an aircraft onto the step in this manner. To become expert requires considerable practice and knowledge of the limitations of the aircraft being handled. It should be emphasized that if the nose refuses to come up progressively higher and the aircraft will not start to rock, it will be impossible to put the aircraft on the step. In order to avoid overheating and causing serious damage to the engine, throttle right back, taxi slowly, and lighten the load or wait for a breeze.

Cross-Wind Take-offs

Provided the wind is not too strong, cross-wind take-offs are entirely practicable. The procedure is identical to that for landplanes. The aileron control is held to the windward side and appropriate rudder pressure applied to maintain the desired direction. When the aircraft leaves the surface of the water, a gentle turn is made into wind, if possible.

Downwind Take-offs

Downwind take-offs are possible and may even be preferred when the wind is light, if obstructions or other circumstances do not favour a take-off into the wind. Hold the control column farther back than when taking off into wind, otherwise the procedure is the same. Much more room is needed for a downwind take-off. In a small body of water completely surrounded by land an excellent procedure is to begin to take-off downwind and finish it into wind. This is done by starting the take-off downwind and when in the planing mode turning into wind, thus bringing the aircraft into the take-off configuration near the downwind shore. A reduction of power may be necessary to avoid upset, or for control during the turn in the planing mode; therefore, do not neglect to apply take-off power again after the aircraft is pointed into wind.

Landings

Landing a floatplane presents some problems that are unfamiliar to the landplane pilot. An airport, except possibly during the winter, always presents the same general surface, whereas the surface of the water is continually changing.

An airport is restricted to air traffic only and is as free as possible of obstructions and obstacles, whereas boats, floating debris, and submerged obstructions are everyday hazards to the floatplane pilot. It is therefore wise to fly all around a proposed water landing area to examine it thoroughly for obstructions such as floating logs and mooring buoys and to note the position and direction of motion of any watercraft.

When a windsock is not present at a landing site, there are several methods for determining wind direction. If there are no strong tides or currents, boats lying at anchor will point into wind. Seagulls and other water fowl invariably land facing the wind. Sails

on boats give a fair approximation of the wind; smoke and flags are other indicators. If the wind has appreciable velocity, its path is shown by streaks on the water, which in a strong wind become distinct white lines. The direction of the wind cannot be determined by these alone, but if there are whitecaps on the waves there is no difficulty. The foam appears to move into wind, an illusion caused by the fact that the waves move under the foam.

Landing Attitudes

Because of the continually changing properties of the water's surface, a pilot can use a wide variety of touchdown attitudes. When the water surface is reasonably smooth, the best touchdown attitude is at an angle such that the steps and the heels of the floats touch at the same time. A nose-high, power-off landing is safe, but it is not as smooth nor as pretty to look at as a "step-heel" landing. It may also be a little disconcerting at first to the landplane pilot as the aircraft rocks forward almost to the level attitude immediately after contact. This is due to the heels of the floats striking the water first, causing a pronounced drag which tips the aircraft abruptly forward. However, remember that a smooth landing may be made in any attitude between the step-heel and nose-high attitude provided the control column is moving steadily back at the instant the aircraft contacts the water and is held back after initial contact to prevent the floats from digging in.

Landing Run

Upon touchdown, the aircraft will slow down through the three taxiing modes in reverse order to take-off — i.e., planing, nose up, and idling. If the landing is made some distance from the docking or mooring point and water conditions are satisfactory, open the throttle sufficiently while the aircraft is in the planing mode and taxi in "on the step." Taxiing in the planing mode is much easier on the engine than taxiing in the nose-up mode. With the latter mode, the engine RPM is the same, but as the forward speed is much lower engine cooling may be inadequate if long distances are involved. In any case, the last few minutes of taxiing should be done at idling RPM to cool the engine and prevent after firing when it is shut down.

Landing in Rough Water

When the waves are high, select the best suitable sheltered location and land with an attitude equivalent to the attitude of the aircraft in slow flight, and plan to touch down on top of the wave. If the aircraft starts to porpoise after touchdown, carry out a missed approach. When landing in rough water, the aircraft will slow down appreciably as it strikes the first wave, but not enough to keep it from slamming into the next one. The shock of this second contact can be lessened by judicious use of throttle during the bounce. Exercise caution while taxiing. This may be an occasion to use nose-up taxiing to turn downwind to proceed to shore or an anchorage.

Glassy Water Landing

Landing on glassy water can be hazardous unless you follow proper procedures. It is impossible to determine the height of the aircraft above the water in glassy water conditions; without special procedures it may be flown into the water or stalled at a considerable height during the "float" after the aircraft has been flared for landing. Either situation can be extremely dangerous.

Power assisted approaches and landings must be used when landing on glassy water. While it may not always be possible, it is desirable to set up a normal approach over the terrain preceding the leeward shoreline and land parallel to a shoreline. If these aids are not available, objects in the water should be used to judge altitude.

When approximately 200 feet above the surface (300 to 400 feet where visual aids for judgement of height are not available) reduce the rate of descent and apply more power. The objective is to produce a safe airspeed and power combination that will result in a nose-up attitude sufficient to prevent the floats from digging in on touchdown. The descent should be established at 200 feet per minute or less by the time the shoreline is crossed. Provided the attitude and airspeed are correct, you need not alter the power until touchdown. Care must be taken to trim the aircraft properly to ensure that there is no slip or skid at the point of contact.

The flight instruments, particularly the airspeed and vertical speed indicators, should be scanned during the final approach, while using peripheral or forward vision for clues of the height above the surface. If the rate of descent increases, increase power until the desired rate of descent is regained while maintaining the correct attitude and airspeed. Make no attempt to round-out or "feel for the surface." At the point of contact, which should be gentle with the steps and heels of the floats touching simultaneously, the throttle should be eased off gently while maintaining the back pressure on the control column to prevent the floats from digging in as the aircraft settles into the water. Remember, considerable space is required for this type of approach and landing.

The approach speed necessary to achieve the cor-

rect attitude and the amount of power used to control the rate of descent will vary with each type of aircraft. Procedures and airspeeds recommended by the aircraft manufacturer must be followed. In the absence of manufacturer's data, the approach speed should be determined by experimentation before you attempt glassy water landings. While figures of 15 to 30 percent above the calibrated stall speed are often quoted for approach speeds, each type of aircraft has to be dealt with individually. If during your seaplane endorsement training no glassy water experience is possible, you must receive dual instruction on the procedure from a qualified instructor before you attempt a glassy water landing as pilot-in-command.

The same landing procedures may be used if failing daylight, deteriorating weather, or other conditions affect depth perception over a landing area.

Should a pilot be forced to land on glassy water after the engine has failed, a landing should be effected as close to the shoreline as possible and parallel to it, the height of the aircraft above the surface being judged from observation of the shore. Floating objects, weeds, and weed beds can also be used to judge height.

Landing in a Cross-Wind

The procedure for landing a floatplane in a cross-wind is much the same as for landplanes. Lower the water rudders as soon as possible after landing but not while in the planing mode. Do not attempt cross-wind landings in high waves. The possibility of one float landing in a trough and the other on a crest could capsize the aircraft.

Downwind Landing

Avoid downwind landings if possible. The high "ground speeds" cause an undesirable forward pitching of the aircraft as the floats make contact with the water. Allow plenty of room and use the soft field technique in the approach and landing so that the floats touch the water at the minimum safe airspeed. Engine power should be left on at point of touchdown only long enough to check the tendency for the aircraft to pitch forward.

Landing at Other than Regular Operating Areas

Before landing, check the intended landing surface and the subsequent proposed take-off path for rocks, sand bars, debris, or other obstructions, and decide on the taxi path and the method of beaching. In marginal areas, it is also wise to select a prominent reference point, from which you can carry out a missed approach if you are not already on the water, or, on take-off, from which you can discontinue the take-off if you are not airborne by that point.

Landing on Land

If it becomes necessary to land a floatplane on land, due to an engine failure or other extreme emergency, plan the approach and execute the subsequent landing so as to contact the ground with the keel of the floats as nearly parallel to the ground as possible. Immediately after touchdown pull the control column hard back.

Float flying can, under normal conditions, be extremely rewarding and pleasurable. Otherwise inaccessible areas are opened up, and provided you pace your learning and confront new situations with respect and discretion, much is to be gained.

Skiplanes

In those areas of Canada that have a reasonably long winter season, an aircraft equipped with skis takes on a versatility as a mode of transportation or sport, that no ordinary landplane has. As soon as the snow cover reaches a depth of 2 to 3 inches, practically every open expanse of flat land becomes a potential landing surface. When the ice on sheltered lakes becomes thick enough, these too offer inviting wide open places for taking off and landing. It is an exciting experience to have take-off and landing areas on all sides practically as far as you can see. It is also a pleasant novelty to be relatively free of the other air traffic normally associated with aerodromes and airports.

Like any other endeavour in the field of aviation, operating an aircraft on skis must be kept within the bounds of certain guidelines. These guidelines require the exercising of basic common sense throughout the many and varied circumstances that may occur in a new and sometimes sensitive operational environment.

In this exercise we will discuss only those operational areas peculiar to skiplanes. The subject aircraft will be a typical tail wheel style, single-engined, light aircraft equipped with fixed skis (as opposed to a retractable ski wheel arrangement). It should be pointed out that certain nose wheel aircraft may be equipped with skis, but most light skiplanes are of the tail wheel style.

Properties of Snow

Studies carried out by various agencies disclose many interesting aspects of the properties of snow and ice, but one point stands out in explaining why a skiplane may perform differently nearly every time there is even the slightest rise or fall of the ambient temperature or change in the texture of the snow. The pressure of the skis as they move on the frozen surface creates an extremely thin film of water between the ski and the surface, even at subzero temperatures. The film of water, depending on the type and temperature of the surface, either lubricates the ski and assists its forward progress or clogs the bottom surface of the ski and impedes it. For example, since a skiplane has no mechanical braking system, if it is landed on the bare ice surface of a lake in even the slightest tail wind, this minute film of water could mean practically no friction between the ski and the surface, so that the aircraft would slide unhampered until it collided with something or some action was taken to redirect its heading. On the other hand, this same film of water between dampish snow and the skis of an aircraft attempting take-off could create a surface tension that would drastically restrict acceleration and lengthen the take-off run considerably.

Manoeuvring on the Ground

A skiplane in flight operates almost exactly like a landplane. The additional weight and the aerodynamic properties of the skis may affect the stalling speed and overall performance of the aircraft, but to a degree that is usually negligible. However, on the ground and especially during the landing roll, the lack of brakes on normal skiplanes calls for special techniques and certain precautions. Unlike the floats of a floatplane, the "heels" of aircraft skis cannot be forced down into the snow or ice by elevator action to arrest forward motion. The steering effectiveness of steerable tail skis fluctuates from good to nil according to the type of snow or ice on which the aircraft is manoeuvring. Even with normally good conditions, the steering effectiveness of the tail ski may deteriorate considerably in a cross-wind. In any case, with no braking system at all, let alone no differential braking capability, sharp taxiing turns are more difficult to execute than with wheeled aircraft. This is especially important in congested areas. In strong cross-winds that would be difficult but not impossible for landplanes to taxi and manoeuvre in, the skiplane may have to depend on outside handling for directional guidance. A great many turns on the ground are done by "blowing" the tail around with bursts of engine power while applying full rudder in the direction of the turn desired. Most of the problems associated with manoeuvring skiplanes on the ground occur in congested areas if there are strong winds. Offsetting this is the fact that skiplanes usually operate away from airports and are seldom required to manoeuvre in congested or confined areas. Almost invariably, the take-off and landing can be made directly into wind (there being no defined runway system); therefore, cross-wind taxiing is usually reduced to a minimum.

Slide

The strength with which skis will adhere to the snow comes as a considerable surprise to most pilots operating a skiplane for the first time. Sometimes even full engine power will not cause the aircraft to move. Between adhesion and the surface tension previously mentioned, the length of the take-off run and landing roll varies greatly with the condition and type of the snow surface. The weight of the aircraft also has more effect on this performance under poor snow conditions than in the same aircraft on wheels, to the extent that it may be necessary to lighten the load in order to get off the ground.

A skiplane attains its best take-off performance when the surface affords as much slide as possible. This is not a profound statement in itself, but it achieves importance when you consider that the surface condition which affords the most ideal landing run, since no brakes are available, could be one that produces as little slide as possible. Bear this in mind when contemplating an operation into and back out of confined space.

The best sliding conditions for skiplane take-offs can occur when the ambient temperature is relatively low or relatively high. At temperatures of −5°C and lower, the film of water caused by compression between the ski and the snow surface is extremely thin, producing good lubrication but very little surface tension. When ambient temperatures rise to the point at which the snow cover becomes watersoaked and slushy on top of a sound surface, lubrication is good and the "smack" of the ski on the slush as it progresses forward breaks the surface tension. Although the sliding qualities of the higher temperature conditions are good, performance itself is not quite as good as at the lower temperatures, due to drag. Drag in this case is caused by the ski sinking slightly into the soft

surface and having to push against the slush ahead of it; in a way it could be likened to the skis trying to ride up over a tiny but everpresent hill. Fig. 2-111 offers a slightly exaggerated illustration of this.

Surprisingly enough, new snow even at fairly low temperatures does not provide a good slide factor, and if deep enough and new enough can affect the take-off performance of a light skiplane very seriously. The reason for this is a combination of drag (Fig. 2-111) and lack of lubrication. New snow contains a great deal of air, which allows the skis to sink in deeply and at the same time acts like a blotter, absorbing the thin layer of water instead of allowing it to lubricate the skis. New snow of reasonable depth on top of older snow provides ideal landing conditions. A thin layer of new snow over ice could be somewhat less than ideal, because of the lack of braking action.

Snow is at its worst for slide when it is at its best for making snowballs. Skiplanes manoeuvring in this type of snow have difficulty making anything but large radius taxiing turns without severe blasting of the tail. The skis tend to stick to the surface, so that a lot of engine power is needed to get the aircraft moving. A peculiarity of skis, however, is that once moving, even at a snail's pace, they will continue to move even in adverse snow conditions. Once the aircraft has broken clear, keep it moving, rather than continuously stopping and starting, to avoid abusing the engine. This type of snow can be misleading since if the aircraft encounters an area compacted by other traffic the slide qualities will suddenly improve. Approach these areas cautiously to avoid unexpected forward progress.

When planning a take-off on new snow, or any snow condition that allows the skis to sink to any depth, compact a take-off path by taxiing the aircraft up and down the proposed take-off area. Do this before loading the aircraft, to avoid overheating the engine and to make the aircraft easier to manoeuvre

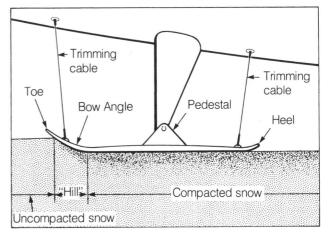

Figure 2-111 Major Components of a Ski

during the compacting process. If for any reason the take-off must be discontinued during the ground run, change the heading of the aircraft so that it leaves the compacted area. This will help bring the aircraft to a stop in a shorter distance, through the aid of drag.

Take-off

The procedures for taking off and landing a skiplane are basically the same as a soft field take-off for a landplane. It is good practice, once the tail of a skiplane has been lifted on take-off, to hold the nose slightly higher than is customary with a landplane. This will allow the toes of the skis to ride up over snow "ripples" instead of digging into them, and will transfer the weight from the skis to the wings as quickly as possible.

On a landplane the tires absorb a great deal of the shock and iron out bumps and other irregularities on the manoeuvring surface. A skiplane has only the suspension system to absorb shock, and it is not generally designed to counteract anything but minor irregularities. You must get used to high noise, vibration, and bumps and shocks not experienced with wheeled aircraft. However, this does not mean that a skiplane has some special quality for accepting shocks that a landplane does not. There are many surfaces that would have to be considered operationally unacceptable for skiplane operations, except in an emergency. Under most circumstances, though, there is reasonable scope for selectivity and you should exploit it fully.

Landing

Most skiplane landings can be made into wind. However, if a landing is made in a cross-wind it must be executed in the usual manner, with the appropriate control action taken to counteract drift. There are many snow and ice surface conditions that can exert more side-load on a skiplane's landing structure, if drift is present at touchdown, than the same amount of drift would exert on a landplane's landing-gear. Here again, the landplane's tire absorbs a great deal of the side-loading. A ski is long and a touchdown with drift in a rutted, icy, or hard-packed snow surface can twist the landing struts beyond the point of return to proper alignment. Ideally a skiplane should be landed in the three-point attitude.

Cross-Wind

It has already been said that because of the large off

airport manoeuvring areas usually available to skiplanes, landings and take-offs can be executed into wind at most times. But on a normal snow surface, which allows the skis to sink into the surface slightly, cross-winds may be handled in the same manner as with landplanes. However, you may have to land a skiplane (or take off) in an area limited in width, such as a river or a narrow serviceable area on a frozen lake. If the ice surface is relatively smooth, the influence of a cross-wind must be counteracted in a manner that is quite contrary to everything said thus far on the subject of cross-wind landings or take-offs. The cross-wind landings and take-off techniques for landplanes (and skiplanes under normal snow conditions) depend largely on a ground surface that will hold the wheels or skis firmly aligned in a desired direction while they are on the surface. A skiplane on ice does not have this advantage. When landing on an icy surface that freely allows lateral movement of the aircraft, a skiplane must come to touchdown by counteracting cross-wind drift in the normal manner. At touchdown the aircraft must be immediately crabbed into wind to maintain the desired direction of travel until it comes to rest. In other words, the aircraft is actually sliding sideways throughout the landing run. More crab is required as speed diminishes. On take-off, apply crab immediately and maintain it throughout the take-off run. Less crab is required as speed increases. Execute this procedure with great care and if possible only with a first-hand knowledge of surface conditions, since hazards exist when crabbing a skiplane on an icy surface in a strong cross-wind. Should the skis encounter a rough spot or a soft area while most of the weight is still on the skis, the aircraft may suffer considerable side-loading. The severity of the side-loading depends on speed, amount of weight still on the skis, and the extent and degree of the rough or soft area. The side-loading would ordinarily be least severe just after touchdown or immediately prior to take-off, since at these points much of the aircraft's weight is on the wings, and the crab angle is smallest.

In normal recreational flying avoid this type of operation, or select an alternative surface that will permit landings and take-offs into the wind with no side drift.

Sticking

The skis of a skiplane parked for any length of time on any snow or ice surface will adhere to the surface. The adhesion will vary in degree, according to weather and other conditions, from one in which reasonable engine power will start the aircraft moving to a condition where no amount of power will get it under way. A skiplane that is stuck fast may sometimes be broken clear by judicious use of simultaneous engine power and movement of the rudder and elevator controls. The wiggle induced will unstick reasonable adhesion, but under no circumstances use this method until you are certain that the tail ski is free. Forward movement of the elevator control may unstick an adhered tail ski so suddenly that it will be almost impossible to prevent the aircraft nose from pitching down and possibly damaging the propeller.

The best way to unstick a skiplane is manually, by the careful use of strategically placed wooden levers. If the skis are so submerged in ice that you must chop them free, make sure that any lumps of ice are removed before attempting to taxi. If the snow condition is such that the aircraft sticks fast each time it stops, unload it and compact the area to be used by taxiing up and down until the surface slide improves.

If the aircraft is to be parked for any length of time over two hours, under most conditions you should taxi it up onto something that will limit the degree of surface adhesion and allow the aircraft to easily break itself clear when it is next used. Spruce boughs, the trunks of small trees, lengths of lumber laid out at right angles to the skis, or well secured plastic sheeting will serve this purpose.

Types of Skis

Aircraft ski construction varies widely, from brass trimmed laminated "boards" to sophisticated hydraulically operated wheel-ski combinations. Most skis are attached to the wheel axle when the wheel is removed, while others are a "roll-on" variety, using the cushioning effect of the tire, with the tire fitting in a channel built into the top of the ski.

The main function of the trimming and restraining cables with which aircraft skis are equipped is to flexibly maintain and secure the skis in a correct attitude during flight. During the walk-around inspection prior to flight include a careful examination of these cables, to ensure that they are in good condition and that all their locking and securing devices are intact. Should a rear cable fail during flight the aircraft would become asymmetrically trimmed and awkward to fly, but if a front cable failed, a very dangerous flight condition and landing problem would exist. The affected ski would drop to approximately a 45 degree angle and create extreme drag on one side of the aircraft. Landing in this condition carries the risk of the toe of the ski digging into the snow upon touchdown, tearing off both the ski and its landing strut. Because these cables are subject to minute inspection by most skiplane pilots, a failure is extremely rare. However,

should a failure occur, search for a landing area with a surface that offers no opportunity for the ski to dig in, such as the snow-free ice on a lake or a river. Use a power assisted approach and landing to keep the nose high. Just before touchdown, enter a forward slip with bank toward the affected ski and gradually allow that ski to settle onto the ice, then remove the slip and complete the landing.

White-out

White-out conditions are most commonly encountered when flying over large lakes at some distance from the shoreline. Never proceed into areas where, due to white-out, it is not possible to distinguish between ground and sky, unless you are sufficiently experienced and capable of instrument flight (see Chapter 6, "White-out").

If it should be necessary to land during these conditions follow the procedure used for glassy water landings (see the section, "Glassy Water Landings," in Exercise 26). The heels of the skis should make contact with the surface first. As soon as this occurs, close the throttle immediately, since unlike floatplanes, the skiplane will have no tendency to pitch forward upon contact with the snow or ice surface. Due to the lack of visual cues the instruments must be used for altitude, airspeed, and descent reference throughout the approach.

Suitability of Snow Surface for Landings

When selecting a landing area under conditions of restricted visibility or bright sun, it is often impossible to determine the condition of the surface. Extreme caution is advised. Ice ridges, hummocks, and windrows of snow are often impossible to see, and any one of these can cause serious damage to a skiplane. The annual toll of aircraft damaged due to selecting unsuitable terrain for a landing is an effective reminder of the need for discretion when deciding to fly under such conditions and when selecting a landing area.

Unlike water surface conditions, there is no relationship between wind conditions and ground conditions. High snow-drifts could exist in a no-wind condition, and snow-drifts could be parallel as well as at an angle to the existing wind.

Thin Ice

When operating skiplanes on ice-covered lakes or other ice areas with a water base, there are various rules of thumb to determine the thickness of ice required to support a certain aircraft weight. However, none of these will safely cover all situations. For complete knowledge on this phase of skiplane operations refer to the Transport Canada publication, *Water/Ice Aerodrome Standards and Recommended Practices*. This publication may be purchased from:

Transport Canada
330 Sparks Street
AARNG
Ottawa, Ontario
K1A 0N8

Operations

Due to the possibility of extending the take-off run, it is important to remove dirt, sod, ice, etc. adhering to the undersurface of the skis before attempting a take-off. When operating off an ice surface remember that ice near the shoreline will bear greater loads in early winter, but should be used with great caution in the late winter and early spring. On rivers, avoid areas of undercurrents resulting in thin ice. Points where creeks or rivers flow into or out of lakes should also be avoided, along with air holes, fissures, and deep snow.

If during a refuelling operation fuel is spilled directly onto the snow or ice surface, move the aircraft a safe distance from the spill area before starting the engine. Gasoline lying on a cold surface does not evaporate as quickly as usual and could be ignited accidentally by the engine exhaust flame.

Survival

While winter operations should automatically include survival equipment, the very nature of ski flying demands extra precautions. Carry ropes, axe, shovel, blowpot, snow-shoes, tent, engine and wing covers, and a receptacle for draining and heating the engine oil, where applicable. Other survival equipment such as food, sleeping-bag, clothing, and footwear should also be considered if you are operating away from home base.

Type Conversion

One side effect of improving the performance of any medium can be increased complexity of operation. The manual change gearboxes in transport trucks and buses afford more efficiency and flexibility than the transmissions in conventional automobiles, but to operate them effectively requires much greater skill than the average car driver normally possesses. And so it is with aircraft. One that requires only basic skill to operate safely usually lacks the overall performance of the more complex.

For our purposes, in addition to flaps, carburettor heat, and mixture control, an advanced-type aircraft in the middle range of complexity is equipped with retractable landing-gear, a constant speed propeller, and a multicell fuel selection system. It invariably has a higher wing loading than an initial trainer; therefore, it also has a higher stalling speed. Extra cockpit equipment will include a propeller pitch control, a landing-gear position indicator, a landing-gear position selector, an engine manifold pressure gauge, and quite often engine cylinder head temperature, exhaust gas temperature, or carburettor air temperature gauges.

Weight and Balance

The weight and balance of a complex aircraft is usually more critical than that of a basic aircraft. The aircraft, with its greater length, fore-and-aft baggage compartments, and greater load-carrying capacity, can easily be improperly loaded. Pay strict attention, therefore, to weight and balance prior to flight.

Complex aircraft, with their greater efficiency of operation, offer the flexibility of carrying a limited payload a great distance with full tanks, or a heavy payload a short distance with limited fuel. This calls for very careful payload and fuel load planning on the part of the pilot-in-command.

Pre-Flight

The pilot's visual inspection of a complex aircraft is similar to the inspection of a basic aircraft. To begin with, when you enter the cockpit to check that magneto and master switches are off, also check that the landing-gear up-down selector is in the "down" position.

When checking the tires and other standard items of the landing-gear, also look for hydraulic fluid leaks in the lines and actuating cylinders of the landing-gear retracting system. If the system is electrically operated, look for loose electric wires and switches. The exposed threads of the screw-type jack of an electromechanical retracting system should be bright and free of accumulations of oil, dirt and sludge.

Complex aircraft often have separate baggage compartments with their own doors. These doors must be securely closed prior to take-off. Besides other obvious consequences, a door that opens in flight and cannot be closed may change the aerodynamic characteristics of the aircraft and produce serious control difficulties. The security of baggage doors is the pilot's responsibility, a responsibility that should never be passed to unqualified persons.

Fuel Systems

When the aircraft has a multicell, multiselection fuel storage system, study the relevant section of the Aircraft Flight Manual very carefully, since there may be a certain fuel tank (or tanks) that cannot be selected for take-off or landing. When so specified there are various technical reasons for this precaution. It is not always the case of course, but generally the fuel tanks that are specified as not to be used for take-off or landing are designated as "auxiliary" and those that may be used as "main." It is good practice to start and warm up the engine on an auxiliary fuel tank to check the system, then select a main tank prior to the engine run-up and take-off. Auxiliary fuel tanks

197

should not be selected until cruising altitude is reached and the aircraft is in a level attitude. The main fuel supply should be selected well prior to landing. Many pilots select the main supply at the beginning of the descent from altitude and then recheck the item during the pre-landing check.

The complex aircraft under discussion is most likely somewhat larger than the initial trainer; therefore, it may be presumed that considerably more fuel is being carried in each wing. In order to keep the aircraft in good lateral trim, the fuel should be used alternately from each side of the aircraft. If, for example, an aircraft is carrying 4 hours of fuel on board (2 hours on each side), fly for 45 minutes on one side initially, switch over to the other side, for 1.5 hours, then return to the original side, which now has 1.25 hours of fuel remaining.

It is never considered good practice to run a fuel tank dry unless a fuel shortage warrants doing so. When a tank is run dry, there is a possibility of sediment or water being drawn into the carburettor or fuel injection system, but more important still, an air lock may be introduced into the fuel system. The best procedure for fuel management when full use must be made of the available supply is to calculate fuel consumption and accurately time each tank, using the fuel gauges as monitors. When the fuel state requires running a tank dry, try to do so at altitude in level flight while at cruising speed. If there is a fuel warning light and it begins to flicker, select a new tank and switch on the boost pump. Be careful not to select an empty tank position. In running a tank dry, if you must wait until the engine falters through lack of fuel, select a new tank and then move the throttle lever back to a position half-way between the idle and normal cruise setting until the engine is running evenly again.

Engine and Propeller

Several principles are used to effect the change in blade pitch angle of the various makes of constant speed propellers, but in the majority of cases engine oil at engine pressure provides the moving force. The propeller must be "exercised" from its high RPM to low RPM range through at least two cycles prior to take-off. This is usually carried out during the warm-up and engine ground check. The main reason for doing this, of course, is to ensure that the propeller mechanism is functioning properly, but it is also for the purpose of circulating warm engine oil through the propeller actuating system.

When the propeller pitch lever is fully forward, the propeller blade pitch angle is at its smallest, and when the lever is pulled right back, the pitch angle is at its greatest. There are various terms to describe the posi-

tion of the propeller pitch lever in relation to the blade angle, but perhaps the most descriptive and least confusing is:
1. Pitch lever forward — high RPM
2. Pitch lever back — low RPM

The engine should be started with the propeller pitch lever in the fully forward position. There may be mechanical reasons for this, but the most important reason is operational. To start the engine of an aircraft in the class under discussion, it must be cranked at about 250 RPM. To crank the engine at this speed the starter must produce a specific amount of power with the propeller pitch at its smallest angle (smallest bite). If the pitch lever is not fully forward, the blade angle will be greater and the starter may not be able to crank the engine fast enough to cause it to start, or if it does start there may be an undue strain on the starter and battery.

When checking the magnetos for drop in RPM during the engine ground check, make sure that the propeller pitch lever is in the "full high" RPM position. Being a constant speed propeller, it will automatically adjust the RPM to compensate for power loss due to a magneto drop if the lever is in some intermediate position. In other words, the drop will be there but the tachometer will not show it.

A constant speed propeller allows the pilot to match the operational performance requirement of an aircraft more readily with its available engine power at a certain throttle setting. Most internal combustion engines attain their maximum horsepower at a point somewhere near their maximum allowable RPM, but the power requirement for economical cruising is usually found at considerably less than maximum RPM.

When listing the performance figures in an Aircraft Flight Manual, the manufacturer usually begins the list with the distance, in feet, that the aircraft requires for its take-off run. The ability to accelerate from a stopped to an airborne condition in as short a distance as possible and then climb out initially at a good rate is a prime performance requirement of most aircraft. To accomplish this the engine must be allowed to develop the RPM that produces its maximum power output. This is done by setting the pitch angle of the propeller blades small enough to allow the engine to rotate freely up to the desired RPM. When the pitch angle is small the propeller takes a smaller bite of the air, but it takes many more bites because it is turning faster. This is much like the need to use a lower gear in an automobile for acceleration, or to climb a steep hill at low speed. Once the aircraft is in cruise climb or cruising flight, the air is coming through the propeller at such a speed that if its blades are left at the smaller pitch angle, it cannot take a sufficient number of small bites to efficiently chew its way forward. (This is like leaving an automobile in low gear while trying to maintain highway speed.) At this point the propeller pitch angle must be increased so that a bigger bite of air can be taken for each engine revolution.

Thus, to use the available horsepower of an aircraft engine effectively, for a given flight configuration, you must be able to control the pitch angle of the propeller blades. The fixed pitch propeller used on an initial training aircraft is designed to provide the best compromise between take-off performance and cruising speed.

An internal combustion engine can be operationally damaged, even to the point of failure, by sustained overspeeding and/or subjection to high internal pressures for lengthy periods of time. To use a familiar parallel again, when an automobile is forced up a steep hill at relatively low speed in high gear with the accelerator depressed to the floor, severe and damaging pressures are being generated within the combustion chamber of the engine. This condition could be remedied by selecting a lower gear. Because it is air-cooled, an aircraft engine is even more susceptible to damage by similar engine abuse. In an aircraft with a fixed pitch propeller, the pilot cannot, under normal circumstances, subject the engine to abnormal internal pressures nor is it likely that engine overspeeding will be allowed to occur for too long a time. This is not so in the case of aircraft equipped with constant speed propellers. The pilot has control over internal engine pressures and can also overspeed the engine very easily.

To assist in maintaining acceptable engine pressures, in addition to the familiar tachometer the complex aircraft is equipped with an instrument called a manifold pressure gauge and perhaps cylinder head or exhaust gas temperature gauges, or both. The engine handling procedure for take-off and climb in most initial training aircraft is to apply full engine power and maintain this power setting until the altitude selected for cruising is reached. To do this with most constant speed propeller equipped aircraft would subject the engine to severe and unnecessary abuse. If the engine is not turbo-charged, take off with full throttle and the propeller pitch control fully forward in the high RPM position. As soon as the aircraft is established in the normal climb attitude at a safe altitude, throttle back to reduce the power until the recommended climb pressure is indicated on the manifold pressure gauge. Then bring the propeller pitch control back until the recommended engine RPM for the climb is indicated on the tachometer. Aircraft with turbo-charged engines often do not use full throttle for take-off. Use the recommended procedures, manifold pressures and RPM shown in the Aircraft Flight Manual for the aircraft being flown.

When you are climbing an aircraft with a constant speed propeller at the recommended manifold pressure and engine RPM, the manifold pressure will decrease as the aircraft ascends. This will require constant forward adjustments to the throttle to maintain the correct manifold pressure. If the climb is continued, you will reach an altitude where the throttle is fully open and the manifold pressure cannot be maintained.

When cruising altitude is reached, bring the throttle back to the manifold pressure recommended for the speed desired at the selected altitude. Then bring the propeller pitch control back until the corresponding recommended engine RPM is indicated on the tachometer. On descent from altitude, you will have to constantly adjust the throttle backward to maintain a desired manifold pressure.

When the pilot selects a specific RPM setting, the governing mechanism of a constant speed propeller will maintain it regardless of reasonable variations in throttle setting and aircraft attitude. With this type of propeller, reasonable movement of the throttle will change readings on the manifold pressure gauge, but will not alter the engine RPM. However, movement of the propeller pitch lever affects both RPM and manifold pressure. When the pitch lever is brought back, the RPM will decrease and the manifold pressure will increase. Conversely, ease forward on the pitch lever to increase the RPM and the manifold pressure will decrease. An increase of manifold pressure by movement of the pitch lever alone is undesirable. Therefore, remember the rule: to increase engine power, first increase RPM; to decrease engine power, first decrease manifold pressure.

To increase power: 1. pitch forward;
                       2. throttle forward;
To decrease power: 1. throttle back;
                       2. pitch back.

Carburettor icing will be indicated in the case of an aircraft with a fixed pitch propeller by a decrease in RPM. This is not the case in aircraft with constant speed propellers. The propeller will continue to maintain its RPM setting until the carburettor is so choked with ice the engine will not have the power to maintain RPM, even at the smallest propeller blade pitch angle. The instrument that will indicate that carburettor ice is present is the manifold pressure gauge. If the altitude and the throttle setting have been constantly maintained, a decrease in manifold pressure may indicate that carburettor ice is present. We say "may" since other factors can cause fluctuations in manifold pressure. The manifold pressure gauge is a pressure instrument and is influenced by the atmosphere; therefore, flight from a high pressure area into a low one would cause the manifold pressure to decrease, and vice versa. A decrease in manifold pressure may also indicate an engine malfunction. In any case apply carburettor heat in the manner prescribed by the manufacturer. If there is a momentary decrease and then a rise in manifold pressure, carburettor ice is most likely present.

You will recall that early in training you were warned against abrupt applications of power, and adequate reasons for such a warning were given. In the case of turbo-charged engines there is another important reason. An abrupt movement of the throttle, particularly on a cold engine can cause an overboost that could exceed the engine limitations. Although the turbo-charger is designed and constructed to accept some abuse, changes in power applied to a turbo-charged engine must be progressive and smooth with the degree of rapidity warranted by the situation.

It is considered good practice to place the propeller pitch control into the "full high" RPM position somewhere in the traffic circuit, prior to landing. Primarily, this assures that full power is available should a go around be required with both landing gear and flaps extended. Some operators leave the pitch control at the normal cruise setting throughout most of the circuit and then place the pitch lever into the "full high" RPM position (forward) after turning onto the final approach. This ensures that adequate engine power is immediately available.

Retractable Landing-Gear

Retractable landing-gear gives the recurring advantage of a gain in airspeed, due to a reduction in parasite drag. For all practical applications, the ratio of the advantage increases as airspeed increases.

An aircraft with its landing-gear retracted has a shallower glide angle than a similar aircraft with fixed landing-gear. For example, should a forced landing become necessary, the gear may be left in the retracted position to gain glide distance and extended at the right moment to reduce speed or increase the glide angle.

Another distinct advantage of retractable landing-gear is that it extends the safe operating speed between stalling and normal cruising speed. This gives you a wider choice of speeds for fuel economy.

Although there are aircraft with manually operated landing-gear, most are operated hydraulically, electrically, or by a mixture of both, with an emergency method of extending the gear manually. Since the systems are remarkably trouble free, pilots are apt to ignore the procedure for manually extending the landing-gear, resulting in delay and confusion should a landing-gear emergency arise. It is good practice to allow a few seconds to review the procedure for manual extension of landing-gear, as part of the pre-taxi geographical check of the cockpit.

When the outside air temperature is below freezing, refrain from splashing through puddles of water or accumulations of slush to avoid the possibility of the landing-gear freezing in the "up" position when retracted after take-off. If it is suspected that the landing-gear is wet or slush covered, delay retraction for a short period after take-off or cycle the landing-gear to reduce the possibility of retraction or extension problems.

Most aircraft manufacturers publish airspeed restrictions concerning the raising and lowering of the landing-gear of a specific aircraft. There often is a maximum airspeed above which the landing-gear should not be retracted. In some aircraft, the gear will not retract above this airspeed. There are also aerodynamic and structural reasons for these speed restrictions. Therefore, it is important to commit these airspeeds to memory.

Under normal circumstances, an aircraft with retractable landing-gear will climb faster than its counterpart with fixed landing-gear. Therefore, in the case of short field take-offs, especially over obstacles, retraction should occur as soon as possible and consistent with safety. Some Aircraft Flight Manuals recommend a delay in this procedure, as the drag of gear-door opening and the retraction process may exceed the drag when the gear is down. Aircraft with devices that protect against accidental retraction of the landing-gear may require special procedures for retraction below a certain airspeed. Before contemplating a short field take-off procedure, review the subject carefully in the Aircraft Flight Manual.

It is not good enough to presume that the actual position of the landing-gear automatically corresponds with the position of the "up-down" selector. An aircraft may be flown for miles while the pilot searches for the reason for poor performance, eventually to find that the landing-gear is still extended due to a popped circuit breaker or a blown fuse. The landing-gear position lights should be rechecked when power is reduced for the climb or as part of the after take-off check. Nearly all systems incorporate an arrangement of green lights to indicate that the gear is down and locked. Red lights warn that the landing-gear is unsafe and they remain on when the gear is not down and locked. During the training period, the landing-gear will most likely be lowered during the cockpit check on the downwind leg prior to landing. However, there may be advantages to lowering it at a point more suitable to the operation. Therefore, it is important to set up a procedure and treat this as an independent or additional check item, to be carried out on final approach. The green and red gear position indicator lights in some aircraft are automatically dimmed when the position lights are switched on. This has caused more than one pilot to believe there was no "gear down" indication because the cockpit was not sufficiently dark to show that the lights were on. Switching the position lights off momentarily will confirm that the lights are functioning. If a light does not illuminate as expected, it may be burned out. Disconnecting a light that is functioning, and inserting it in the place of the suspected defective light will remove all doubt.

General

The handling and performance characteristics of a complex aircraft may differ from those of a basic aircraft as follows:

1. The take-off run may be longer.
2. Greater torque and thrust may require coarser rudder movements to keep the aircraft straight during the initial stages of the take-off run.
3. It is more difficult to regain control if directional control is lost during the take-off run.
4. There may be poorer forward visibility in the climb attitude.
5. Climb will be faster, requiring more vigilance for slower aircraft ahead or in the vicinity of the airport traffic circuit.
6. Stalling speed is higher.
7. The radius of a standard rate turn is usually larger.
8. The majority of complex aircraft have a low wing; therefore, other aircraft at a lower altitude are obscured at certain angles of vision.
9. A substantial reduction in airspeed occurs when the landing-gear is lowered.
10. More foot pressure may be required to operate the rudder of nose wheel aircraft when the landing-gear is lowered.
11. The approach and landing speeds are higher.
12. The combination of increased weight and higher landing speed may lengthen the landing roll.

Amateur-built and Ultra-light Aeroplanes

In the last decade there has been an increase in the number of amateur-built and ultra-light aeroplanes. These aircraft are built from plans or kits developed by knowledgeable designers or engineers. However, these aircraft have not been subjected to the rigorous flight test standards required for formal certification. In the case of an amateur-built aeroplane, the original builder had to establish a limited performance envelope during its first 25 hours of flight. Ultra-light aeroplane builders may or may not have developed flight performance criteria for their particular aircraft.

The majority of these aircraft are well built and are meticulously maintained. However, because of the lack of standardized design, construction, and maintenance, pilots should consider the following points before they fly an amateur-built or ultra-light aeroplane:

1. Find an instructor or pilot experienced on the aircraft you intend to fly. Each one has unique flight characteristics. It is estimated that as many as half the accidents involving ultra-light aeroplanes involve pilots trained in certified aircraft. Finding a pilot experienced in the aircraft you intend to fly will save training time and could prevent an accident.
2. Flight control inputs may dramatically differ from what you are used to in a conventional aeroplane.
3. The ground handling characteristics may be different than you expect.
4. Instruments, if they are installed, may not look the same as they do in a conventional aeroplane.
5. Automobile engines are sometimes utilized in amateur-built aeroplanes. Make yourself aware of the engine type and its unique operating characteristics.
6. Two-stroke engines are also a possibility. Two characteristics are worthy of note. First, carburettor heat is rare. Second, due to their design, two stroke engines do not have the built in redundancies of certified aircraft engines. Therefore, engine performance should be carefully monitored.
7. Safety features built into certified aircraft may or may not be incorporated into an amateur-built aeroplane.
8. The stall characteristics of the aircraft may not have been evaluated.
9. In an ultra-light aeroplane, the difference between the stall speed and the normal climb speed may be very small. If the engine fails during a climb, most ultra-light aeroplanes will decelerate rapidly. Quick recognition and reaction is crucial.
10. Some amateur-built aeroplanes may be authorized for aerobatics. This is on the basis of an owner's declaration, not a flight test evaluation by an experienced test pilot.

If you are a novice, seek out experienced advice so that your flight will be enjoyable and safe.

Emergency Procedures

Although aircraft today are very safe, there are times when a malfunction could adversely affect the safety of the flight. During your training you will learn the proper methods of dealing with these problems.

As emergencies rarely happen in well-maintained aircraft, their occurrence is usually unexpected. A pilot who is not mentally prepared for an unfamiliar situation, may take inappropriate actions to deal with the problem. Being prepared by having a pre-determined plan of action will help you make good decisions and also reduce the possibility of making a distraction induced error. Therefore, it is important to periodically review standard emergency procedures in your Aircraft Flight Manual.

In almost all situations the aircraft will be capable of controlled flight. However, the distraction, preoccupation, and channelized attention caused by an emergency, or unusual event such as a door, window or panel opening in flight, may significantly degrade a pilot's control of the aircraft. Your primary task is to fly the aeroplane while you assess and deal with the problem.

Some emergencies require a fast response. In these situations you must have the vital actions for the aeroplane you are flying committed to memory. The check-list may be used to follow-up as time permits. The emergency procedures check-list in most Aircraft Flight Manuals provides easy reference for immediate corrective action and follow-up. Indecision costs time and in some cases time may be crucial.

Many emergencies don't happen quickly, but are the result of a series of events. If the pilot is alert to subtle changes in a situation, the chain of events that leads to an emergency may be broken.

It is very difficult during training to cover every emergency situation that you might possibly experience. Understanding the operation of your aircraft's systems will help you to cope with unusual situations. By considering such emergencies while on the ground, you will be better prepared if one occurs in flight.

Some typical emergency situations are discussed in the following paragraphs.

Engine Fires on the Ground

Most engine fires occur due to improper starting techniques often caused by lack of or non-adherence to a check-list. Incorrect starting procedures for the time of year can also lead to engine fires. For example, during cold weather a pilot may overprime an engine that refuses to start. This action often leads to backfiring, which could cause a carburettor fire if the engine does not start. Overpriming is also common in summer weather where, in many cases, little or no priming is required.

The procedures for dealing with an engine fire on the ground are usually well detailed in the appropriate Aircraft Flight Manual and should be committed to memory. Know the recommended starting procedures, the manufacturer's recommended drill for starting problems, and the steps to follow in case of an engine fire.

When the vital actions for fire are complete, depending on the urgency of the situation, you might consider using the radio to request assistance before evacuating the aircraft. *DO NOT* fly the aeroplane after the fire has been extinguished as there may be internal damage of which you are unaware. Have an Aircraft Maintenance Engineer inspect it before flight.

Fires in Flight

Most Aircraft Flight Manuals include procedures to follow in the event of a fire in flight. Very specific procedures for engine, cabin, and wing fires are sometimes outlined separately. These should be committed to memory and reviewed periodically.

Activation of the fire extinguisher may become necessary to extinguish a cabin fire. It is worth a word of warning that this may cause temporary oxygen depletion in the closed space of a light aircraft cabin or fill the cabin with a fine powder that restricts visibility. The cabin should be ventilated immediately according to the manufacturer's instructions.

In rare instances fires occur that cannot be extinguished; this requires an *immediate* landing.

Electrical Fire

Modern aircraft rely more and more on electrical systems. Even light aircraft are not immune to electrical fires and most Aircraft Flight Manuals give clear procedures on how to deal with this type of emergency.

The initial steps in most procedures call for turning off the master or battery switch and the alternator or generator switches. The purpose of this is to remove all power to the systems to prevent further shorting while you attempt to identify and isolate the faulty system or unit. It is important that careful consideration be given to the effect that turning off these switches will have on lighting, communication, and navigation systems that you might be depending on at the time.

When attempting to isolate the cause of the fire, only turn on equipment that is absolutely necessary and turn each item on one at a time with a significant waiting period between each item. Do not rush; with electrical fires it often takes time for the malfunctioning electrical component to heat up and start smouldering again.

Should the pilot turn the systems on too quickly and a burning smell is noticed, the tendency will be to turn off the last unit activated. That unit *may not* be the one causing the fire.

If a circuit breaker has popped or a fuse has blown it is likely associated with the fire. The Aircraft Flight Manual should be consulted before resetting a popped circuit breaker or replacing a fuse.

Some electrical fires generate a considerable amount of acrid smoke. Should this occur, follow the Aircraft Flight Manual's smoke removal procedures for a cabin fire.

Electrical fires can cause a high degree of anxiety and the isolation procedure can be time consuming. Remember, one of your primary concerns is to fly the airplane. Do not become so distracted with procedures that you let a more serious situation develop.

Icing

VFR pilots should not be flying in icing conditions. Extreme caution must be exercised when planning a flight in conditions even remotely conducive to icing.

Icing does not occur only in cloud. You could be flying well below the cloud base in relatively good visibility and encounter freezing rain. The hazardous effects caused by ice accumulation on your aircraft in flight requires an immediate emergency response. Should an inadvertent encounter with icing occur, usually the best procedure is to note your heading and turn 180 degrees to get back into clear air. Follow the emergency procedures for icing outlined in the Aircraft Flight Manual. In particular, you should ensure the pitot heat is on, if available, and get as much heat as possible to the windshield. Inadvertent flight into icing conditions can be most effectively avoided by careful pre-flight planning and attention to existing and forecast weather conditions.

Carburettor Icing

Carburettor icing can seriously affect the safe operation of an aircraft. Its formation can be subtle, causing a smooth, steady, power reduction: or it can be quite evident with rough running and rapid loss of power.

Carburettor icing may cause the engine to stop. Therefore, it is important that carburettor ice be detected early. Being aware of the potential for carburettor icing on a particular day, watching and listening for signs of engine power loss or roughness, and periodically applying carburettor heat are effective ways to prevent serious carburettor icing.

The procedures for using carburettor heat to prevent and eliminate carburettor icing are outlined in the appropriate Aircraft Flight Manual. It should be noted that the application of carburettor heat when carburettor ice is present may result in an increase in engine roughness and a further decrease in engine RPM. This is normal and is caused by the melting ice and water passing through the engine. Therefore, leave the carburettor heat on until the engine operation smooths out.

Electrical Problems

Electrical power supply system malfunctions in most light aircraft fall under two categories. They are *excessive rate of charge* and *insufficient rate of charge*.

The onset of these malfunctions are silent and can lead to an emergency if not immediately detected and countered. Therefore, it is imperative that you monitor the electrical charging system continuously and systematically throughout the flight.

Excessive rate of charge.   If a sustained high battery charge is noted, a higher than normal voltage in the electrical system is possible. This can adversely affect aircraft electrical components and may, if not dealt with immediately, cause the battery to overheat and evaporate the electrolyte at an excessive rate. The appropriate checklist in your Aircraft Flight Manual will outline the procedures to follow for this type of malfunction.

Insufficient rate of charge. Some modern aircraft have a warning light included as part of the electrical system. When this light illuminates it acts as a warning that there is a problem with the alternator and it has been shut down. The battery is now no longer being charged. If this occurs, follow the procedures as outlined in the Aircraft Flight Manual. Generally, they include a shut-down of all unnecessary electrics to conserve battery power. If the aircraft does not have a warning light, no charge or a discharge will be indicated on the ammeter. It is wise to inform the appropriate ATC unit of the problem in the event that radio contact is lost.

Aircraft equipped with generators rely on relatively high engine RPM for maximum output. Generator output should be checked during run-up to ensure an adequate rate of charge. If a discharge is noted during flight, the same procedure as described in the paragraph above should be followed.

Low Oil Pressure

Although an indication of low oil pressure may warrant immediate attention, shutting down an engine without further investigation could result in an unnecessary forced landing, or a forced landing into less than favourable terrain.

Trouble-shooting this problem should start by checking the oil temperature. If the oil temperature indication is steady and within limits the problem may be a faulty oil pressure gauge. However, if oil temperature is rising, an engine failure may be imminent. The specific procedures for dealing with low oil pressure in flight may vary, and it is important that the appropriate Aircraft Flight Manual be consulted as soon as possible. Remember, a thorough pre-flight inspection will usually prevent low oil pressure emergencies.

In-Flight Panel or Door Opening.

Cabin and baggage door latches and panels on most light aircraft do not provide a conspicuous visual indication that they are not properly secured, nor are such aircraft usually equipped with a warning device

in the cockpit. Therefore, pilots should carefully check both the condition and security of all panel and door latches before flight to ensure they do not open on take-off or in flight.

Should a panel or door open on take-off or in flight, in almost all situations the aircraft will be capable of controlled flight. However, during take-off or initial climb when a pilot's workload is high, the distraction, pre-occupation, and channelized attention caused by an inadvertent opening may significantly degrade a pilot's control of the aircraft. **Remember, your primary task is to fly the aeroplane** while you assess and deal with the problem.

As the flight characteristics of aircraft with open panels or doors vary by type, consult the Aircraft Flight Manual for specific guidance on handling of inadvertent panel or door openings on take-off or in flight, and on door re-closing procedures.

Ditching

The possibility of a single-engine landplane having to make an emergency landing in open water is remote. However, should there be no other choice, follow the procedures in the Aircraft Flight Manual. In the absence of manufacturer's data, there are general procedures that may be applied.

Decide as early as possible that ditching is inevitable, so that power can be used to achieve a stabilized approach at minimum rate of descent and low airspeed consistent with safe handling.

Reports on slow, fixed-gear aircraft that have been ditched indicate that the main gear tends to dig in during the initial impact and prevents the aircraft from skipping and subsequently striking the water in a stalled, nose-low attitude. The aircraft simply decelerates rapidly and the nose burrows only slightly. However, because aircraft with retractable landing-gear generally have higher landing speeds and are subject to greater deceleration forces, when able, the landing-gear should be retracted.

Low wing aircraft should be landed either with flaps retracted or extended only slightly. However, full flap should be used on high wing aircraft unless otherwise specified in the Aircraft Flight Manual.

Consideration should be given to cracking the door and wedging it open or to opening the canopy to reduce the possibility of jamming. Fasten seat-belts, secure loose objects, and ensure that all equipment needed for flotation and the prevention of hypothermia is available at hand.

Attempt to determine the wind speed and direction by observing the surface conditions. Waves generally move downwind except close to the shoreline, but swell does not bear any relation to wind direction. Wind lanes (streaks upon the water) may be apparent, the streaked effect being more pronounced when looking downwind. Gusts may ripple a smooth surface and indicate the wind direction. From the air water appears to be calmer than it really is. If possible, fly low over the water and study its surface. When near the surface, the aircraft's drift should give a good indication of surface wind conditions and direction.

Land into wind if the water is smooth, or smooth with a very long swell. In some situations such as large waves or swell, or swells with short spacing, it may be advisable to land parallel to the swell and across the wind. The danger of nosing into large waves or swell is greater than the danger of landing cross-wind. When ditching on a river, unless a strong wind dictates otherwise the landing should be made downstream to reduce impact speed.

Be prepared for a double impact. The second and greater impact will occur when the nose hits the water. The aircraft may also swing violently to one side if one wing touches the water before the other. Release seat-belts when certain the aircraft has stopped and evacuate the cabin as quickly as possible. If the doors cannot be opened immediately, do not panic. It may be necessary for a considerable amount of water to enter the cabin before the pressure equalizes and the doors can be opened. Unless it is badly damaged, the aircraft will not sink immediately.

Conclusion

With a little care and planning many of the emergency situations mentioned can be prevented. Proper maintenance of the aeroplane and a thorough pre-flight inspection will go a long way toward reducing the chance of a problem. Knowledge of how your aircraft systems work will give you an advantage when it comes to making decisions concerning emergencies. Prevention is also enhanced by maintaining good situational awareness, managing the risks as things change, avoiding hazardous attitudes, making sound decisions, and adhering to a check-list.

Radio Communication

Clear, concise, and accurate radio communication between the pilot and a ground station is essential to flight safety. Use the Standard Radio Telephony procedures and phraseologies outlined in the A.I.P. Canada and in the *Radiotelephone Operator's Handbook*. Pay close attention to all instructions and clearances issued from any Air Traffic Control unit to prevent misinterpretation and to ensure you receive and understand all relevant information. When in doubt, request clarification.

You may at first find that you are unable to understand many transmissions and the situational awareness they impart, but it will come together with experience. One method used by instructors to develop a student's "ear" for transmissions is to have them listen to professional pilots on a radio with an aircraft radio frequency band. It can also be used to receive airport information on what is known as an ATIS transmission.

Automatic Terminal Information Service (ATIS)

ATIS is the continuous broadcasting of recorded information for arriving and departing aircraft at major airports. Its purpose is to improve controller effectiveness and to relieve frequency congestion by repetitive transmission of essential but routine information. ATIS for a particular airport is assigned a discrete VHF frequency or may be heard over specified navigation aid frequencies.

Each recording is identified by a phonetic alphabet code letter beginning with "ALPHA," and as conditions require a change in the message, subsequent letters of the alphabet are used. Once the ATIS message has been heard, the pilot should inform the ATC unit on first contact that the information has been received by repeating the code word identifying the message. ATIS frequencies are listed in the *Canada Flight Supplement*.

Clearance Delivery

Major airports often have a Clearance Delivery frequency assigned, which ATC uses to reduce radio congestion on other channels. This frequency is used primarily by ATC to issue Instrument Flight Rules (IFR) clearances to aircraft on the ground. However, in some cases at these airports all departing aircraft are required to make contact after receiving ATIS. Upon initial contact with Clearance Delivery, the pilot should state the aircraft type, registration, and ATIS message code received.

After Clearance Delivery has responded to the initial call, the pilot's intentions should be stated. After all the necessary information has been received by both the pilot and controller, the pilot is usually told to contact ground control. The Clearance Delivery controller passes the information obtained to the next controller to be contacted, normally ground control.

Consult the *Canada Flight Supplement* to determine if a call is required on Clearance Delivery and the assigned frequency.

Ground Control

At most controlled airports, taxi clearance is normally received through contact with ground control. The method of contact and information passing is described in the A.I.P. Canada. If there is no ATIS available, the ground controller will ensure that all pertinent information is relayed to the pilot. The ground controller is responsible for clearing the pilot to and from the active runway and that clearance often includes certain instructions that must be followed.

Upon receipt of a normal taxi clearance, a pilot is expected to proceed to the taxiway holding position for the runway assigned for take-off. No further clearance is required to cross any non-active runway en route. However, *under no circumstances* may a pilot taxi onto an active runway unless specifically cleared to do so. Should the ground controller require a pilot to request further clearance before crossing or entering any taxiways or runways while en route to the holding point, this will be stated in the taxi instructions.

To assist in preventing active runway incursions, taxi instructions that contain the word **HOLD** shall be acknowledged by a read back of the hold point by the pilot.

Examples of hold points that should be read back are:

HOLD or **HOLD ON** (runway number, taxiway);
HOLD (direction) **OF** (runway number); and
HOLD SHORT OF (runway number, taxiway).

Furthermore, with the increased simultaneous use of more than one runway at many airports, instructions to enter, cross, backtrack or line-up on any runway should also be acknowledged by a read back.

Tower Control

Departing

The tower controller is responsible for the orderly flow of traffic departing from and landing at airports, as well as all VFR aircraft flying within the airport control zone.

When ready for take-off, or when instructed to do so, the pilot should switch to the assigned tower frequency. Often this frequency is very busy; therefore, pilots must be very careful not to transmit until they are certain that all other radio transmissions between the tower controller and other aircraft are completed. Do not request take-off clearance until all pre-take-off checks have been completed, because once the tower has given the clearance there should be no undue delay taxiing onto the runway and taking off. If possible, the aircraft should be turned so that the approach may be scanned for other traffic. Even though the airport is controlled, the pilot still has a responsibility to ensure that there is no conflicting traffic when cleared to taxi onto the active runway.

After requesting take-off clearance, the pilot may be told to **HOLD** due to traffic approaching or on the runway. The aircraft must not proceed farther until a new clearance has been received.

The pilot may also receive a clearance to "taxi to position and hold," which means that the aeroplane may be taxied onto the runway and lined up ready to go, but must not take off until cleared. In both of the

above instances, the pilot must read back the hold portion of the clearance.

When "cleared for take-off," the pilot shall acknowledge, taxi onto the runway, and take off without delay, or inform ATC if unable to do so. The tower controller may ask if the pilot is ready for an *immediate* take-off. This means that the pilot is expected to taxi onto the runway and take-off in one continuous movement.

The pilot may have requested a specific altitude or direction after take-off. It cannot be assumed that the request will always be fulfilled. The tower controller may issue different departure instructions than expected, or alternate instructions any time after the aircraft has lifted off. The pilot must continuously monitor the frequency in case there is a change.

IFR traffic may also be flying within the control zone, but could be on other assigned frequencies. The tower controller and IFR controllers communicate with each other to ensure a smooth orderly flow of traffic; however, pilots flying in the area may not be totally aware of all potential conflicting traffic. Thus, the need to keep a sharp look-out for other aircraft cannot be overemphasized.

Once airborne the pilot, unless otherwise cleared to do so, is to remain on tower frequency until clear of the vertical or horizontal boundaries of the control zone. Pilots must always monitor the tower frequency when within a control zone as the controller may wish to pass traffic information or issue instructions. Clearances and instructions to other aircraft may also affect your flight. By keeping a good listening watch on the frequency you maintain situational awareness, which assists in identifying potential traffic conflict.

Arriving

When approaching an airport with the control tower in operation, if available, pilots should obtain the ATIS information well in advance. If possible, the tower frequency should also be selected early. This procedure provides the information necessary for the pilot to plan the most expedient entry into the control zone. Before entering the control zone, the pilot must contact the tower controller and advise whether the intention is to land or proceed through the zone. A clearance must be obtained from the appropriate ATC unit prior to entering a Class "B" or "C" control zone.

Should the intention be to land at the airport, the pilot may expect the shortest routing to the runway in use if traffic permits. Any necessary restrictions will also be included in the clearance.

One such restriction is a holding over a geographic location, or a VFR check point or call up point found on the VFR Terminal Area Chart (VTA). Pilots are expected to do a left-hand orbit within visual contact of the point and be prepared to proceed directly to the airport immediately upon receipt of further clearance. If such a restriction is unacceptable, pilots should inform the controller and state their intentions.

A clearance "to the circuit" by the controller means the pilot is expected to join the circuit on the down-wind leg at circuit height. Depending on the direction of approach and the runway in use, it may be necessary to proceed cross-wind prior to joining the circuit on the downwind leg.

A clearance for a "straight in approach" authorizes the pilot to join the circuit on the final leg without having to fly any other portion of the circuit.

When established mid-downwind the pilot is to advise the tower controller by calling downwind. At this point the pilot may indicate the type of landing intended, such as a full stop or a touch and go, or request "the option." If "cleared for the option," the pilot may make a low approach, a touch and go, a stop and go, or a full stop landing. This procedure is usually used when traffic is light.

Normally, the controller will initiate the landing clearance when the aircraft is on final approach without having first received a request from the pilot. However, should this not occur, the onus is on the pilot to request such clearance. A pilot must obtain a landing clearance prior to landing. If it is not received, the pilot shall, except in an emergency, overshoot and make another circuit. As well, for various reasons, such as traffic still on the runway, the pilot may be told to "pull up and go around." Should the pilot decide that an overshoot is necessary, the tower controller should be advised as soon as possible.

After landing, the tower controller may issue instructions to exit the runway, but the controller does not usually expect an acknowledgement from a pilot who is still busy with the landing roll. If no instructions are received, the pilot is expected to continue to taxi in the landing direction to the nearest suitable taxiway and exit without delay. Normally the aircraft should be taxied forward to a point at least 200 feet from the runway or across the hold line before coming to a stop. When off the active runway, taxi instructions will be given by the ground controller.

Transiting a Control Zone

Pilots who intend to transit a control zone should follow the initial procedures for arriving aircraft and state their intentions. The controller may then provide altitude and routing instructions or vectors to fly and local traffic information. In some instances, such as heavy traffic or poor weather, or both, entry to a control zone may be refused. Depending on the reason,

the pilot must then be prepared to hold, divert, or in the case of weather perhaps request special VFR.

Vectors

During flight training you will become familiar with terms involving direction, such as magnetic heading, track, compass heading, etc. But when an ATC unit requests that you fly on a certain heading, do not become involved in any calculations; simply fly your aircraft on the requested heading as indicated by the magnetic compass or the correctly set heading indicator. Most likely ATC is observing you on radar and will be providing guidance by manoeuvring the aircraft by means of vectors. (*Vector*, simply stated, is another way of saying heading.)

Advisory Service (Uncontrolled Aerodromes)

Civil aerodromes that do not have control towers are called uncontrolled aerodromes. These sites may be operated by municipalities, corporations, individuals, or Transport Canada. At many uncontrolled aerodromes Flight Service Stations (FSS) offer an advisory service to aircraft operating into and out of the aerodrome or in the area. Flight Service Stations may provide services to a remote aerodrome through a Remote Aerodrome Advisory Service (RAAS). A departing aircraft initiating a call to a Flight Service Station might receive the following reply:

GOLF CHARLIE VICTOR HOTEL / TIMMINS RADIO / WIND 040 AT 10 / ALTIMETER 3011 / WIND IS FAVOURING RUNWAY 03 / AIR ONTARIO FLIGHT 166 REPORTS ON APPROACH TO RUNWAY 03 / NO OTHER REPORTED TRAFFIC / TAXI AND TAKE OFF AT YOUR DISCRETION.

As a Flight Service Station does not exercise control of air traffic, an aircraft receiving the above message is free to taxi and take-off provided correct radio procedures are followed and no traffic conflict exists. Transport Canada has designated a Mandatory Frequency for use at selected uncontrolled aerodromes, normally those served by Flight Service Stations, Remote Flight Service Stations and Community Aerodrome Radio Stations. Some airports that are uncontrolled at times when the control towers are closed are also assigned a Mandatory Frequency. Aircraft operating on the ground or in the air within the area in which a Mandatory Frequency has been designated must be equipped with a functioning radio capable of maintaining two-way communication.

The Mandatory Frequency will normally be the

frequency of the ground station that provides the advisory services for the aerodrome. The specific frequency, distance, and altitude within which the pilot must comply with Mandatory Frequency procedures is published in the *Canada Flight Supplement*.

An Aerodrome Traffic Frequency is usually designated for active uncontrolled aerodromes that do not meet the criteria for a Mandatory Frequency. The Aerodrome Traffic Frequency is established to ensure that all radio-equipped aircraft operating on the ground or within a specified area use a common frequency and follow common reporting procedures. The Aerodrome Traffic Frequency is normally that of the ground station where one exists or 123.2 MHz where a ground station does not exist. The specific frequency, distance, and altitude within which use of the Aerodrome Traffic Frequency is required is also published in the *Canada Flight Supplement*.

More information concerning the use of Mandatory and Aerodrome Traffic Frequencies can be found in the A.I.P. Canada.

VHF Direction Finding System (VDF)

A pilot who is unsure of the aircraft's position and is able to contact a VHF/DF equipped control tower or FSS, on frequencies listed in the *Canada Flight Supplement*, may be provided with headings to reach the VDF site. This site is normally located on or near an airport. The pilot may also obtain a bearing from the VDF site, track out assistance, and an estimated time or distance from the site. As well, a fix may be obtained when used in conjunction with another VDF site, a VOR radial, or a bearing from an NDB. Details of this service are outlined in the A.I.P. Canada.

Epilogue

The preceding pages of this manual have outlined background knowledge and techniques that have been formulated through the years by the learning experiences of others. It is impossible to put in print solutions to every possible situation that might arise, but the common sense application of the messages that this manual has tried to transmit should leave you well prepared to face the world of aviation. Receipt of a new licence or rating is an acknowledgement that you have reached a certain level of competency and knowledge — you now have a licence to learn. Exercise the new privileges you have earned with care and responsibility. Keep an open mind, and do not be afraid to ask questions or question the validity of ideas that have been accepted for many years as being gospel. Only a questioning mind will find more effective solutions to problems and record information that can be passed on to future generations of pilots by means of manuals such as this. Those of us who have put this manual together are fully aware of the responsibility we bear and appreciate the opportunity we have been provided to pass on to you, the reader, the lessons learned from thousands of hours of collective experience as flight instructors.